THE
APE
THAT
SPOKE

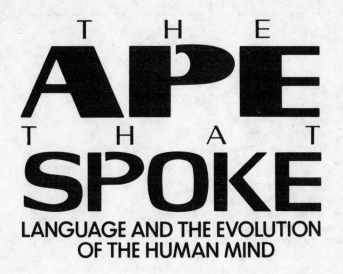

THE APE THAT SPOKE

LANGUAGE AND THE EVOLUTION OF THE HUMAN MIND

JOHN McCRONE

AVON BOOKS ◆ NEW YORK

AVON BOOKS
A division of
The Hearst Corporation
1350 Avenue of the Americas
New York, New York 10019

Published in hardcover by William Morrow and Company, Inc.; for information
address Permissions Department, William Morrow and Company, Inc., 1350 Avenue of
the Americas, New York, New York 10019.

The William Morrow and Company edition contains the following Library of Congress
Cataloging in Publication Data:

McCrone, John.
    The ape that spoke : language and the evolution of the human mind / John McCrone.
        p.   cm.
    Includes bibliographical references.
    1. Human evolution.   2. Brain—Evolution.   3. Language and languages—Origin.
4. Self-perception.   5. Consciousness.   I. Title.
GN281.4.M4   1990   90-48753
573.2—dc20   CIP

First Avon Books Trade Printing: April 1992

AVON TRADEMARK REG. U.S. PAT. OFF. AND IN OTHER COUNTRIES, MARCA REGISTRADA,
HECHO EN U.S.A.

Printed in the U.S.A.

OPM   10 9 8 7 6 5 4 3 2 1

Contents

Author's Warning	7
Understanding the Mind	9
1.	The Balloon-Headed Ape	15
2.	The Fisherman's Net	49
3.	Rousing Memories	85
4.	Thinking Aloud	124
5.	From Baby Talk to Strong Language	143
6.	Strange Voices in the Head	176
7.	Pure Emotions and Romantic Notions	210
8.	Watching the Watcher	234
9.	Truly Self-conscious Man	258
Bibliographical Notes	264
Index	284

Author's Warning

Books about the mind have a poor track record. They are usually either mystical meanderings or jargon-ridden textbooks. More seriously, they often lack a unifying viewpoint to bring the subject into focus.

This book tries to view the mind from a single illuminating perspective. It is based on the assumption that the human mind must have evolved; that self-consciousness must have a biological basis. It then uses plain language to take the reader through some difficult territory: the origins of language, the evolution of habits of thought, and the "mapping" of the world in the brain which creates awareness. It ends up with the controversial conclusion that the human mind is only a few degrees different from an animal's and that self-consciousness, memory, and higher emotions are all simple language-driven abilities which we pick up as children.

To make the text read as clearly as possible, I have done things that would normally be frowned upon. Instead of breaking up the flow of the words by quoting sources as I go along, or stopping to spell out where other people's speculation ends and mine starts, I have written the book as a seamless whole. I trust I have made up for this in the notes at the end of the book. These show where the supporting information came from, which interpretations are my own, and where there are conflicting opinions. And having just insulted everyone else who has written a book on the mind, I mention some of the excellent books that are in existence.

A word about terminology: A lot of everyday words used in talking about the mind are fairly poorly defined. To avoid confusion, I give here a brief glossary of the common terms.

Consciousness   The sensation of being aware of the world around us.

Self-consciousness, self-awareness   The extra step of being aware of what is taking place inside our own heads.

Memory   A general term that covers both natural abilities of the brain, such as the ability to recognize, and learned abilities, such as the ability to recall and replay events in our lives.

Emotion   Another general term that covers both "raw" feelings such as arousal or thirst and human-invented or "higher" emotions such as patriotism and guilt.

Thoughts   Anything that passes through the mind, be it words or images, but usually referring to an attempt at problem-solving or commentary on life.

Understanding the Mind

It all started with an ape that learned to speak. Man's hominid ancestors were doing well enough, even though the world had slipped into the cold grip of the ice ages. They had solved a few key problems that had held back the other branches of the ape family, such as how to find enough food to feed their rather oversized brains. Then man's ancestors happened on the trick of language. Suddenly, a whole new mental landscape opened up. Man became self-aware and self-possessed. He broke free of the grip of the present—the moment-to-moment life lived by all other animals—and became master of his own memory. Language allowed man to relive his past, plan for the future, and step back to consider the fact of his own existence. Through speaking, man rapidly developed a self-conscious mind.

That, in a nutshell, is the story of the human mind. The discovery of speech sparked a revolution in man's mental abilities. But to see how language could have arisen, to see how it could have led to control over memory and how it could have given us self-awareness, we will have to take a more leisurely tour of the evolutionary story behind the mind.

To many people, the human mind seems like some magician's trick. Lift the bony dome of a person's skull and what do we find? No more than a pinkish wrinkled mush. Yet hidden in this pink blob exists a world of shimmering panoramas and cascading thoughts. Inside every head we know that there is a thinking, feeling mind that mirrors our own. It is a puzzle that a surgeon can chop through a human brain and find no trace of the inner mental world we know is there. We might be certain that an answer to it all lies buried in the racing electrical patterns and oozing chemicals found in the brain. But this still does not seem to be a satisfactory reply to the question of how the mind can exist. The mind still seems like nature's most marvelous conjuring trick.

Even our own minds are hard to focus on clearly. We can see roughly the basic parts of our mind. There is the broad sweep of all life's sights and sounds dancing inside our skulls. Then alongside this—but living a paler existence—is the private world of our rememberings and imaginings. And then, intruding into the welter of impressions, are occasional sharp surges of emotion that well up from some dark depth into conscious experience. Finally, almost buried by this swelling tide of impressions and feelings, we can hear the constant chatter of an inner voice. This is the faint voice that appears to rise up inside our heads from nowhere. It often surprises us with the forceful intelligence with which it speaks—although this inner voice often seems simply to echo the passing parade of life, like some rambling TV sports commentator offering opinion and analysis on the scenes our eyes witness.

We can number the parts of the mind, yet we cannot put them under the microscope. When we try to pin down a passing thought or fleeting emotion to examine it properly, it becomes hazy and slips out of our grasp. Details that seemed solid from a distance slide out of sight as we reach out for them.

This book attempts to make clearer how the brain can come to create the ghostly inner world inside our heads. It also hopes to reveal the surprising extent to which our cultural inheritance—the great baggage of learning and social ideas every child picks up—is responsible for the many special qualities that set the human mind apart from an animal's. It should show too how language is the tool with which all the mind's special abilities have been fashioned.

It will do so by following two guiding principles. The first is always to look at the mind as an active process—something that the brain does—rather than as an object with an existence that is somehow separate from the brain. The second is to follow the evolutionary story of the mind.

In the past, philosophers in particular have muddied the waters by treating the mind as if it were an object with a separate existence, independent from the flesh and blood of the brain. They have talked about the mind as a phantom spirit tucked away in a corner of the brain, supposedly fed a picture of the world by our senses, and then, jerking on ghostly strings like a puppet master, somehow managing to make the body dance to its bidding. This confused view of the mind should disappear once we realize that the word *mind* is simply a convenient label for describing the brain at work. The brain can be doing many things at any particular moment, carrying out actions like seeing, thinking, imagining—and even being self-conscious. When we take all these different actions together, we label the resulting mixture "the mind." But, speaking correctly, we never have two separate objects—the brain and the mind—occupying the space within our skulls. We have just the brain and the host of things it can do.

The traditional misuse of the word *mind* can lead us into many wrong assumptions. As well as mistaking the mind for some phantom object separate from the flesh and blood workings of the brain, we make a similar error in talking about the different parts of the mind. We refer to memory rather than the act of remembering, to thought instead of the act of thinking, to the inner voice instead of the act of speaking silently to oneself—and even to self-consciousness rather than the act of thinking about our own thoughts.

Unfortunately, this habit of putting abstract labels on actions is so deeply embedded in our everyday language that it has been impossible to write this book without resorting to doing the same thing. For the sake of readability, words like *consciousness* and *self-awareness* are frequently used when what is really being talked about is the raw sensing of life and the act of reflecting on these same sensations. But by being clear about the pitfalls of language where it matters most, we should be able to explore the workings of the human mind without being left at the end of the day with the unsatisfactory feeling that its essence still somehow eludes us. If we understand how the brain works, then that is all there is to understand. There will be no phantom creature called *mind* that has burrowed deep into the darkest recesses of the brain, escaping our scrutiny as we peel back the layers.

The second guiding principle of the book is to follow the evolutionary story of the mind. It may be an obvious statement, but the human mind must have evolved. It cannot have sprung fully formed from nowhere, turning a dull-witted ape into a glowingly self-conscious human being. The mind must have been shaped over time by the same evolutionary pressures that made man walk on two legs or gave him a delicately gripping hand. By following the evolutionary path that led to modern man, we should get a better idea of how the human mind came together.

One of the biggest evolutionary puzzles is not that humans are self-conscious, but that we could have become self-conscious with such phenomenal speed. It took about a billion years for the primitive brain of our vertebrate ancestors to evolve into the reasonably intelligent brain of an ape. Yet within the space of about two million years—a mere eye blink in geological terms—self-aware man sprang into being. A couple of million years might sound a long time, but to see it in perspective, it is as if the human race had spent all day walking just to travel as far as the ape brain, then reached the rational, self-aware human mind in just three minutes' extra journey.

This tells us that the brain of man cannot be physically very much different from that of an ape—or indeed any other mammal. There has simply not been enough time for evolution to make major changes in the way our brains work. Of course, the human brain has swollen to about three times its original size during the past couple of million years, which seems to be a fairly major change but in fact is rather a trivial difference. It may give us more of what we could term "raw" intelligence, but this is still not enough to account for the apparently huge mental gulf that we feel exists between man and other animals. After all, there are plenty of mammals, such as elephants, whales, and walruses, with bigger brains than ours, so size is not everything. What did make the difference was the arrival of language.

When our ancestors first learned to speak, they unwittingly discovered a tool to reshape their animal minds. They were able quickly to find new ways of using the raw brain given to them by billions of years of painstaking evolution.

Before language, the brains of all animals were driven by the demands of the world around them and so were strictly tied to the present moment. Animal brains could only react—even when they were reacting as intelligently as a chimpanzee's brain. But by a stroke of enormous luck, early man managed to evolve language, and once established, language quickly bred new habits of thought that allowed man mentally to break free of his surroundings. Man became an active thinker about life rather than a passive reactor to the events of the world around him.

Armed with the twin principles of evolution and an emphasis on action, we will try in the following pages to lay bare the workings of the mind. We shall tackle the most puzzling features of the mind, from photographic memory to the inner voice, from the origins of language to the mechanisms of self-awareness. By the end of the story, we will once again look inward into our own minds and this time we should not see a cloudy mess of thoughts, feelings, and impressions. Instead, equipped with an understanding of how the mind evolved, we should be able to see the sharply etched outlines of the mind of an ape that learned to speak.

ONE

The Balloon-Headed Ape

Evolution is essentially a simple affair. The classic definition is that animals give birth to a varied bunch of offspring, the strongest of which live to breed again, while the weakest die off. Evolution is a game of statistics. Those individuals with the right characteristics to survive and breed soon outnumber the less able.

The human mind can only have flowered because it fitted in with this evolutionary game. Each step toward the brain of modern man must have helped that generation survive and breed. But before looking in detail at the evolution of the brain, we should first have a picture of how the apes evolved as a whole.

The story starts about twenty million years ago when the ape family began to flourish. For about ten million years, the world enjoyed a hot, wet era known as the Miocene, when thick forests covered much of the world. Under these ideal conditions, the ape family was much bigger, with many more species than the five lines that survive today—gibbons, chimps, orangutans, gorillas, and man. The success of the ape family was relatively brief, for after the ten-million-year era, the weather changed again. The earth slowly became cooler and drier, causing the vast forests that so suited the apes to retreat and be replaced by open grassy plains and scrubland.

The tree-swinging apes faced a serious problem. They had made fairly major changes to their upper bodies to haul themselves through the branches of the Miocene forests. Their shoulders had been thrown back square and their arms had grown long, strong, and supple. Such limbs were fine for tree swinging but not for getting around on the ground, and as the forests gave way to rolling open plains, it was a question of adapt or die.

Many of the Miocene ape species perished in Africa, where the climate change was most dramatic; the ancestors of the chimps and the gorillas managed to survive by hitting on an efficient compromise. To get about on the ground they balled up their fists and bowled along on their knuckles. This allowed them to scamper quickly over short distances, but such an awkward gait could never match the efficiency of four-legged walking, and chimps and gorillas had to stay on the scrubby forest fringes where they could forage on the ground but scurry back to the safety of the trees if danger threatened.

In Asia, some of the rich Miocene forests survived, and there the slender gibbon and hairy orangutan could continue living a life in the trees. Even so, the development of a tree-swinging body looked as if it had been a mistake. The apes seemed to have swung themselves up an evolutionary blind alley. Having enjoyed a brief ten million years of glory when conditions were favorable, they found that with a slight shift in climate they were too specialized to turn back to a more efficient four-legged gait. At the same time, other evolutionary problems were starting to catch up with the apes. They were getting too big, too slow-breeding, and—paradoxically—too intelligent for their own good.

At first sight, it would seem that intelligence is a good thing and animals could never have too much of it. But brains are very expensive to run. The human brain seethes with the activity of billions of tightly packed nerve cells, which are constantly hungry and, astonishingly perhaps, use up energy as fast as do hard-working muscles. So energy-sapping is the human brain that although it accounts for only about a fiftieth of the body's weight, it burns up about a fifth of the oxygen taken in by the lungs. Being conscious is actually very hard work.

The expense of the brain is an important point. Once we appreciate that our brains are burning up so much of our energy to create consciousness, the common assumption that the mind is somehow effortlessly produced by the brain is destroyed and it becomes easier to accept that the mind is not some ghostly resident in the head. We are also made to realize how difficult it must have been for evolution to justify giving us such a hungry organ. Evolution could never afford to waste energy on human intellect unless it paid off on the bottom line—success at breeding—which it obviously did.

For the apes emerging from the Miocene, the size of their brains was, however, getting to be a problem. With the apes' lush forests rapidly disappearing, a big brain could become more of a millstone that an advantage. Several whole Miocene families quickly vanished. And today, four of the five surviving lines are verging on extinction. The pygmy chimpanzee and mountain gorilla of the African rain forests number only in the hundreds, while the lowland gorilla, orangutan, and gibbon species all number in the tens of thousands. Even the common chimp, which numbers about 180,000, has barely enough members to populate the suburb of a human city.

If it had not been for a twist in the evolutionary story, the apes would probably have been one of the briefer chapters, but fortunately, the fifth line of apes did find the answer to the quickly retreating forests. It was this ape, the distant ancestor of man, who first stood up straight to stride out onto the grassy plains of the late Miocene.

Walking was not as easy an option as it might seem. For a start, it required another bout of body changes hard on the heels of the evolution of tree-swinging arms. More important, two feet cannot move as quickly as four and early man would have been exposed to the big cats prowling the African veldt.

It is often thought that man's ancestors stood upright to free their hands for using weapons and tools, but the first hominids or upright "ape-men" appeared about four million years ago—several million years before the great explosion in intelligence and use of tools took place. The first hominids had brains little larger than any other ape and the only immediate advantage that walking gave them was that they could trek miles across the landscape each day in search of food and shelter. A two-legged gait may not have been fast, but it was a lot more efficient than the shuffling knuckle-running to which chimps and gorillas resorted. If hands made free by walking were an advantage, it was probably because nimble fingers could pluck seed from the rolling acres of grass that spread across the open plains, rather than because they could swing a weapon or tool.

For the hominids to survive their early years on the exposed plains, they must have depended upon having close-knit social groups, compensating for their vulnerability by being alert and organized. Social organization is certainly a hallmark of both monkeys and apes. It was what allowed monkeys, like baboons, successfully to join the hominids on the plains. Despite being four-legged and equipped with big fangs, baboons are tempting prey to leopards and other grassland hunters, but baboons stick together when out in the open, with the males on the edges of the group ready to chase off predators. Many eyes are watching for trouble and many teeth are ready to drive off an attack from a leopard.

Whatever the combination of factors that allowed early hominids to survive on the plains, they moved in a relatively short time from an awkward chimplike shuffle to a flowing stride. The lower half of the body changed in many small ways to allow an upright stance. The bones in the foot fused and the legs lengthened. The thighbones sloped together until they almost touched at the knee, giving a knock-kneed look and allowing hominids to put one foot efficiently in front of the other. The hips and pelvis swiveled around to cup the guts and provide better attachment points for the buttock muscles. The spine took on the snaky curve of a shock absorber and the head was lifted upright to balance out its weight. In next to no time, the first hominids had human hips and legs to match the human arms and shoulders gained in the tree-swinging period. All they lacked was a human head.

This is quite the opposite of the Victorian image of early man. Fossil diggers of the last century pictured our forebears as big-brained apes shuffling along in a half-crouch like a chimp. In fact, the reverse was the case. The first hominids had human legs and bodies but small apelike heads. Although they had slightly larger brains than the average ape, the real increases in brain size did not take place until long after the hominids had made a success of walking.

Indeed, even when hominids had got this close to being human, it was still possible that the human tripling in brain

size that was to come might never have happened. Of the
several closely related species of ape-men living several million years ago, only one line flourished and went on to produce modern man. The other hominid species died out fairly
soon, proving to be unsuccessful experiments of nature, so
that being large, upright-walking, and reasonably bright was
not an automatic recipe for evolutionary stardom. Man might
well have become merely another footnote in the story of
evolution, along with the gorillas, chimps, and Asian apes.
What were the extra ingredients that made man's ancestors
different?

Scientists have only a sketchy picture of the several species of hominids that lived on the African plains about three
to four million years ago. All the bones dug up so far would
fill only a few shoe boxes and most of them are shards of leg
bone or pitted teeth. However, two main families appear to
have existed, each with a number of species. One line of
hominids was slimly built and quickly developed a weak jaw
yet large brain. The other was sturdy with a relatively small
brain and a great grinding set of teeth. This second family—
called *Australopithecus* or "southern ape"—had some members with jaws so massive that they developed a bony ridge
like a Mohican crest running across the top of the skull, which
was needed just to anchor the jaw muscles. It appears that
Australopithecus became a hardy vegetarian, specialized in
chewing very tough roots and coarse leaves, and found its
niche on the dry plains because of its ability to chomp through
the most unappetizing menu.

The slimmer ape—*Homo*—had by contrast a diet that
gradually led to a weaker jaw and smaller teeth. Man's ancestor must have become specialized as a nimble eater of berries, fruit, insects, shellfish, birds' eggs, and small game.
Eventually *Homo* learned to cook and hardly needed teeth at
all—but that is later in the story.

Both ways of life were fine for a while and the two families of hominids appear to have lived side by side. With different diets, they were not in competition, but then the sturdy vegetarian *Australopithecus* went into dramatic decline, most probably due to the arrival of the ice age. The cool drier weather that pushed the forests into retreat at the end of the Miocene was just the slow start to a drastic change in the earth's climate, and after about five or six million years of gradual cooling, the world was rocked by a sudden descent into the real ice age.

The effect of an ice age is dramatic. It does not just ice up the poles but drops temperatures everywhere around the world by about ten degrees centigrade. The world's wildlife gets squeezed into a band near the equator and even here life is hardly comfortable. The vast polar ice packs lock up a lot of the earth's water, disrupting rainfall and turning previously lush tropical areas into drought-stricken deserts.

During the four billion years of the earth's history, the planet has been gripped several times by ice ages. The most recent ice age started about three million years ago and has been particularly severe. It is not even finished yet. Brief ten-thousand- and twenty-thousand-year periods of warmer times have been followed by one-hundred-thousand-year periods back in the grip of the glaciers, a cycle that has so far been repeated about ten times. We are at present living through one of the ten-thousand- to twenty-thousand-year respites between big freezes and so can expect the glaciers to return again soon—that is, if human pollution and the greenhouse effect have not overthrown the climate.

The alternation of ice ages and thaws has put a tremendous strain on all modern-day animal families. Many species, like the woolly mammoths, cave bears, and saber-toothed tigers, adapted to the periods of cold by putting on shaggy coats and growing large bodies to conserve heat, but the regular interruptions of brief thaws have led to explosive comebacks by warmer-climate animals and the extinction of the ice-age specialists.

The severe climate changes of the ice ages may well have been too much for early man's vegetarian cousin, *Australo-pithecus*. Either he had become too specialized with his diet and massive jaws to keep adapting to the changing weather or else greater competition from other animals during the hard times proved too much. Lightweight *Homo* with his flexible diet and brighter mind would have been better suited to such a constantly changing world. In fact, the hardship of the ice ages would have fast pushed him to become even more flexible and even more intelligent, accelerating the trend. Whatever the reasons, all the *Australopithecus* species had vanished by about one and a half million years ago, leaving only the *Homo* line to cope with the bitter winters and brief thaws.

Having dashed through the evolutionary story of the bipedal apes, we can go back and follow at a more leisurely pace the rise of intelligence, culture, and consciousness in man. We shall set the clock back about four million years to when the first small-brained but upright hominids stepped out onto the African plains.

Bipedalism may have been a forced move for man's ances-tors, but it opened many new doors, helping with the solu-tion of at least one other key problem that had been hampering the progress of the apes—that of being able to breed fast enough.

Biologists talk about animals as having one of two types of reproductive strategies, classifying them as either K- or r-type breeders. An r-type animal produces as many offspring as possible and relies on the chance that some will survive to maturity. It does not waste energy looking after its young but instead puts all its resources into producing as many as it can. An oyster is an extreme example of this since it can spawn 500 million youngsters a year, almost all of which will get eaten within a few weeks. Among the mammals, the rabbit is a classic r strategist since it can have twelve infants a year, of which only a few will live to breed themselves.

K strategists, on the other hand, have fewer offspring but take time to look after them, so the young are more likely to survive into adulthood. The K approach demands a greater level of social organization and so usually a greater intelligence in the parents. In the fish world, K strategists do not simply drop tens of thousands of eggs into the water and then forget about them. The mouth breeders popular with tropical aquarium keepers, for example, guard the eggs in a hollow until they hatch and then shelter the young fry in their cavernous mouths.

Mammals, as a group, have taken a great step up in K strategy over their reptile forebears, keeping offspring safely inside their bodies until ready to "hatch" and then feeding them with milk until they are old enough to fend for themselves. Among mammals, the apes are the branch that has taken K to its greatest extreme. While even a mammal as big and intelligent as a lioness produces about two cubs each year, apes give birth to a single infant only once every three to five years. A chimp mother very rarely has twins and nurses her baby for three or four years. Given that female chimps do not normally become pregnant before the age of ten, most chimps cannot raise more than half a dozen youngsters at most during their lifetime.

This exceptionally slow breeding rate pushes a K strategy dangerously close to the limit. The chimps—and the gorillas and orangutans—may have evolved the intelligence and social order needed to make sure that nearly all their children

will survive to maturity, but a price has to be paid in vulnerability to natural disasters, diseases, and habitat changes. If
something happens to wipe out half a chimp troop, then, unlike r strategists such as rabbits and dogs, the chimps cannot
bounce back to full strength during the next summer's breeding season.

So regardless of the challenge of Africa's open plains, the
apes already appeared to face a serious evolutionary problem.
They were on an upward spiral of greater intelligence and
sociability that threatened in the long term to leave them
wide open to swift extinction. However, when the hominids
became two-legged walkers, they seem to have found a way
out of the K-strategy blind alley by discovering how to rear
more than one infant at a time. Human mothers can have
babies in comparatively rapid succession, and rather than each
child being spaced out by at least four years, the gap can be
as little as a year.

If humans bred as slowly as chimps, the results would be
dramatic. Human parents need to look after their children
for about twice as long as chimps: A chimp infant is fairly
independent at about four while a human child might just
about be able to fend for itself at eight. Even taking into account that humans live longer, if they bred at the rate of the
apes, human mothers might still manage only two or three
children in a lifetime. Yet it is common for human families
untroubled by food shortages or overpopulation worries to
have six to a dozen children. Clearly, humans have somehow
managed to push the K strategy even further than the apes
without paying the penalty of lower birthrates. We have the
fruits of the K strategy, such as greater intelligence and a
longer period of care for children, but we have also managed
to rear much larger families.

A key question in our story of consciousness is just how
humans managed this feat. An alien zoologist landing on our

planet about four million years ago would have seen the K
problem and reasoned that the apes had become about as intelligent as they were ever likely to be. Further increases in
costly brain power would only be accompanied by another
drop in birthrates in a game of diminishing returns. Yet
somewhere along the line, the hominids broke the mold. They
were able to push birthrates back up while also tripling their
brain size.

The basic change was the obvious one of the overlapping
of children. Unlike apes, human mothers stopped waiting for
their first child to grow up before having the second—or third,
fourth, and fifth. For this to happen, a number of major social
changes were needed, such as food sharing, the use of campsites, the division of labor between the sexes, a change to a
higher-quality diet, an extended family, and stable "marriages."

Bipedalism probably helped in opening the door to some
of these social changes. Having free hands made it possible
to carry helpless infants or food across the plains back to the
camp. Also, the very limitations of bipedalism were probably
important in shaping man's social behavior: Because hominids could not outrun other dangerous animals, they had to
become more socially organized to defend themselves from
leopards and hyenas.

How and when man's helpful new social patterns became
established is not easy to say since, unlike physical changes
such as upright walking, there is no direct fossil evidence of
them. The only evidence is circumstantial. But it is likely
that many of the social changes either happened as bipedalism was developing—the two types of change reinforcing each
other—or arose soon after.

To see how these social changes fit into the picture, it helps
to look at the most primitive way of life we can see today—
that of a wandering tribe of hunter-gatherers. Man and his

hominid ancestors have been hunter-gatherers for most of their four-million-year history. It is only during the past ten thousand years that man has turned to growing crops and herding animals. If we are tuned by evolution to a particular way of life, then it is to roam in small bands across the countryside, living off the land as we go. Most existing hunter-gatherers, like the !Kung of Africa's Kalahari Desert, the Congo pygmies, and Australian aborigines, live in groups of twenty-five to thirty-five people. They need to move about a lot because even a rich landscape can usually support only about two people to the square mile compared with the hundreds that can live off a square mile of farmed land.

Just as early scientists mistakenly thought of early *Homo* as a large-brained ape, they also wrongly saw him as a savage killer. Even into the 1950s, man's ancestors were being described in lurid terms. The celebrated South African fossil collector Raymond Dart called them "carnivorous creatures that seized living quarries by violence, battered them to death, tore apart their broken bodies, slaking their ravenous thirst with the hot blood of victims and greedily devouring livid writhing flesh." Since then, modern hunter-gatherer societies have been accepted as a more likely model, although perhaps scientists now go too far, making early man sound as if he lived like bands of peace-loving hippies: wandering the plains, picking tasty mushrooms, and snaring the odd rabbit. Nevertheless, it certainly seems that the most natural way of life is one of gentle food-gathering.

Today's hunter-gatherer tribes get about three quarters of their food by harvesting nuts, fruits, edible roots, shellfish, insects, and eggs. They depend on their intelligence and knowledge of the land to survive. In a drought, for instance, the African !Kung can spot the few shriveled leaves that lead deep below the ground to a juicy root. Gorillas, too, show how important intelligence is to this gathering way of life:

They can identify as many as a hundred different berries, fruits, and leaves that are safe to eat and have a far more varied diet than most vegetarian animals. Early man, with a larger brain, must have been even better able to make the most of the food available in the wild.

Hunter-gatherers also, of course, hunt. As well as spearing big game, they can snare small game, catch fish, and even scavenge from carcasses left by bigger predators. But meat is normally a welcomed addition to the diet of a hunter-gatherer rather than a staple food and most primitive tribes are better pictured as being wise shoppers rather than bloodthirsty hunters.

Generally, humans have an easy time living off the land. Even though the !Kung, who were once called the original affluent society, have been pushed into rather poor country by the spread of civilization, they still managed to gather most of the food they needed in a few hours a day. By contrast, chimp troops spend about half their waking day in a hunt for food and, lower down the scale, baboons do little else but forage.

In looking back four million years to the appearance of the first hominids, we have to discount the sort of advanced technology that eases the lives of even the most backward of modern human societies—advances like language, tools, shelter, clothing, and fire. Early hominids had to base their original hunter-gatherer life-style on far more fundamental social advances, such as food sharing, the division of labor, and cooperation—social advances so basic that their importance is often overlooked.

Food sharing is a behavior almost unique to humans. The !Kung of the Kalahari will pull the smallest tortoise or guinea hen into a dozen pieces when it is taken from the fire so that everyone gets a fair share to eat. Any individual trying to sneak more than his portion may be threatened with being kicked out of the camp, a threat as good as a death sentence when survival in the wild depends on being part of a group.

By contrast, other animals may hunt together—as do prides of lions—but there is no patient sharing out of the spoils once a kill is made. Every animal uses its weight to push in and grab as big a mouthful of the carcass as it can. Male lions often do not take part in the chase, yet still force their way in front to eat, while the young cubs are left to fend for themselves. After the feast, the faces of the pride can be covered with blood from the scratching and biting in the struggle to get at the prey. It is true that food sharing of a sort takes place between many animal mothers and their offspring—for example, a mother leopard and mother starling will both bring food back to feed their young—but such sharing does not take place between adults.

Chimpanzees—our closest living relatives—show some signs of humanlike food sharing, although again this is not obvious among the adults. In the wild, they may live together in stable bands about thirty or forty strong, but when it is time to feed, the troop usually breaks up into small groups to go off and forage. This way the troop can cover more ground, but it is also necessary because if one chimp finds a clump of berries, it can expect a more dominant chimp to come over and snatch them away. Even so, there is usually a special tolerance between a mother chimp and her grown-up children. Because young chimps stay close to their mothers for so long, the natural food-sharing behavior of mother and infant can last long past childhood, and it is quite common for a mother chimp to go off foraging with her two-year-old youngster and a couple of fully grown brothers and sisters.

Another sign of chimpanzee food sharing was seen by biologists after a gang of males had managed to corner and grab a baby baboon. The males were willing to share the carcass with each other and even to allow females and youngsters in the group to take a small share.

However, it is only humans who have taken food sharing to an extreme. Hunter-gatherer tribes routinely share both captured game and all the nuts, roots, and grubs gathered during the day. This is a tremendous advance, for youngsters are certain of being well fed during the long years between weaning and becoming mature enough to fend for themselves. Even more important, a group of humans can split up for the day, with some men going off on long-distance hunting trips, while the women, tied by their many children, stay in the relative safety of the home camp and do the gathering of food. The hunters can return knowing there will be something left to eat if they have been unsuccessful, while the women and children can be sure the hunter will trek back many miles to share his spoils with them. By a division of labor, tribes of early man would have been able to make the most efficient use of their energy while giving the maximum protection to their children out on the exposed African plains.

But food sharing did not come automatically to humans. It demanded a strong sense of social structure. An anthropologist living among the Congo pygmies noted that the pygmies were as fair about sharing as any other primitive tribe: A deer captured in nets that the pygmies had strung out in the jungle was painstakingly shared out. However, when an individual pygmy speared a smaller animal on a lone trip, rather than share it as custom demanded, the pygmy tried to get the anthropologist to hide it away in his hut. The presence of an outsider tempted the pygmy to ask for something he would find too dangerous to ask of another tribesman.

Sharing is clearly not inborn in humans. It is just a very useful way of behaving that became established with early man and was one of the keys to his great success. Another key social change that took place somewhere along the line

between the first ape-men and humans was pair bonding—
the development of stable, loving, adult couples.

Monkeys and apes show a great variety of family patterns.
Adult relationships run from the strictly monogamous marmosets to the so-called tournament species, such as baboons,
where the top male runs a harem of females but does little
to support his many offspring. Man's closest relatives, the
gorillas and chimpanzees, both tend to be tournament species. Troop leaders dominate other males with their strength,
thus getting most of the chances to mate with the females,
and the male chimps and gorillas then do little to help the
females in child care or feeding. Clearly this limits the number of infants a troop can support.

Early hominids probably had the same tournament-type
group organization as chimps and gorillas, but evolution soon
pushed them toward the other end of the relationship spectrum, with stable couples sharing the burden of child rearing.
As evidence of this, various physical changes took place in
early man which are typical of monogamy. Like most monogamous animals—and unlike the other African apes—the
size of both sexes became very similar. Human males are
only about 15 percent heavier than females, while male gorillas can be about double the weight. The extra size is needed
in tournament species so that males can be sure of gaining
physical dominance. Another change in humans was that females gradually lost the blatant signals, such as a chimp's
swollen rump, that showed they were fertile and ready to
mate. Instead, females became sexually active—and attractive—most of the month around. They dropped the periodic
"come on" to all males in favor of a closer and full-time attachment to one male.

These physical changes show how swiftly evolution moved
to fashion hominids for stable monogamous partnerships, although the move toward monogamy was not, of course, absolute. There are enough examples of adultery and multiple

wives to show that man has only become relatively more monogamous than his ape relatives. Nevertheless, the shift had important effects on the ability of early man to raise more children. Stable pairing would have cut down dramatically on rivalry between male hominids and so have allowed them to cooperate in hunting and other group activities. It would also have encouraged them to help the mothers feed the off-spring. Along with the other big social advances of food shar-ing and a division of labor, early man would have had the richer food supply and protective family unit needed to raise more than one child at a time. Through these social changes, early man would have been able to break out of the old K trap.

The more organized and cooperative early man became, the faster he could populate the African plains. More food coming in meant he could also start supporting children with larger brains, who in turn could grow up to be capable of ever smarter ways of finding food. The reason for a tripling in brain capacity in hominids in a short few million years was simply so they could breed like rabbits. The bottom line with evo-lution is always reproductive success.

Although it is difficult to pinpoint exactly when all the social advances took place in hominid history, we can look at the many little physical changes that make humans so apparently different from the other apes, and see how evolu-tion came up with such a massive increase in the size of the human brain.

The first point is that humans are in fact exceptionally closely related to the other African apes, the gorillas and the chimpanzees. Until recently, it was believed the African apes and man split off from a common ancestor about twenty mil-lion years ago, but research now suggests that only four mil-lion years or so probably separate the two-legged hominids and the gorilla and chimp lines.

A measure of the closeness of the African apes and humans is the number of genes they have in common. Most well-established families of animals share about 90 percent of the same genes; in the salamander family, for example, the remaining 10 percent of unique genes are enough to account for all the variations in size, color, and physiology seen in different salamander species. Yet with humans and chimpanzees, nearly 99 percent of the genes are identical; only a single percent separates us. It is only our biased viewpoint that makes us classify the apes—and even earlier hominids like *Australopithecus*—as separate families rather than the closest of cousins.

Nonetheless, while the genes have changed little, the appearance of humans has changed greatly. If we imagine looking at the changes on a film where the fast-forward button has been pressed, we would see a hairy small-headed "ape-man" melt into a smooth-skinned human. The head would bulge alarmingly, like a balloon, the face would squash flat, then sprout a pointed nose and jutting chin, body hair would fall away, the arms would thin out and shorten, while the legs would straighten up and lengthen.

Yet, in fact, all that happened during the four million years was a tinkering with growth rates. The bridge of the nose grew longer, the body hair thinned out, the jawbones grew fuller, the skull became larger. The proportions of the parts of the body may have changed, but all the original hominid bones and organs are still there.

The genes that control the growth of the body have "on/off" switches that are easy for evolution to adjust. For humans to evolve longer legs, for example, the timer on the gene switch has simply to be reset so that the leg bones keep growing for another few years during childhood before the gene is turned off. The work of gene switches can be clearly seen in the way we have transformed dogs through selective breeding in the past five hundred years. Hitting the on/off buttons at the right times during a puppy's development can leave it with short legs or long legs, long muzzle or squashed-up face, floppy ears or pointed ears, curly hair or bristles, as small as a chihuahua or as big as a Great Dane. In the same way, evolution could quickly adjust the length of early man's limbs, the spread of his body hair, and the angles of his joints. The major genetic changes that make up the 1 percent measurable differences from the chimps were probably the early ones needed for upright walking. The rest could be left to simple growth-rate adjustments.

The human face, for example, looks not unlike the flattened and weak-jawed face of a baby ape. If changes in diet meant early man no longer needed the massive teeth, grinding jaw, and protruding lips of apes, all evolution had to do was freeze his facial development at the baby-ape stage. Later, when evolution needed to add modern man's jutting chin and resonant nose chambers—both adaptations to speaking—it just had to reset the gene switches so that the jaw and nose went on growing a bit longer than usual.

Changes in growth rates also explain the development of the brain. When we think that it took hundreds of millions of years of development from the first crawling sea creatures to the era of the apes in the lush, green Miocene, it is obvious that little can have happened to the brain in the mere four million years between ape and man. At most, there can have been a resetting of growth switches so that the ape brain ballooned in capacity. The underlying ape structure must still exist, little changed, which is an important point to remember when considering human powers of consciousness, higher emotion, and rational thought. If the only physical difference between ourselves and the apes is the size of our brains, either we exaggerate our mental difference or we use the brain that we have inherited in a subtly different way.

Nevertheless, it is impressive how much bigger man's brain

is compared with his ape relatives'. The brains of both chimpanzees and orangutans weigh about 12 ounces, which is less
than a quarter of the weight of a modern human's—about
three and a quarter pounds. Fossils of early hominids show
a brain size that is only about 2 ounces larger than the average for apes, ranging between 13 and 16 ounces. After a
million years or so, the brain had swollen to an average of
about 17.5 ounces. Then by two million years ago—about
the time of the first tool-using hominid, *Homo habilis*—brain
capacity had risen to 21 ounces or about double the average
for apes.

Such marked brain growth in one branch of a family of
animals is quite rare. Usually, the same evolutionary pressures and constraints affect all the species in a family in much
the same way so that the whole group steadily becomes more
intelligent together. But the growth of the human brain did
not stop at the doubling in size of the first tool-users. Once
tools started being used, growth became even more rapid. The
next important race of early man, *Homo erectus*, had a 35-
ounce brain and by the time modern man or *Homo sapiens*
replaced *Homo erectus* about forty thousand years ago, brain
weight had increased to 52.5 ounces, over four times that of
the original ape ancestors.

Big brains are not, however, a sure route to evolutionary
success since, as we have seen, the expense of large brains
can drag a species' breeding rates down to dangerously low
levels. Man's close cousins, the vegetarian *Australopithecus*,
had brains ranging up to about 19 ounces, which would have
made them smarter than chimps or gorillas and nearly as smart
as the first tool-using *Homo* with whom they shared the African plains about two million years ago. Yet that did not
stop *Australopithecus* from becoming extinct. For man's
ancestors suddenly to become so massively brainy, the payoff in successful breeding rates must have been exceptional

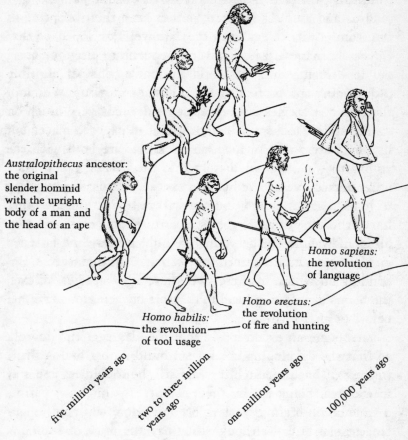

Australopithecus species: heavy-jawed vegetarian offshoot of the hominid family, which finally died out about one million years ago

Australopithecus ancestor: the original slender hominid with the upright body of a man and the head of an ape

Homo sapiens: the revolution of language

Homo habilis: the revolution of tool usage

Homo erectus: the revolution of fire and hunting

five million years ago

two to three million years ago

one million years ago

100,000 years ago

Early man's family tree with the important advances made along the way indeed. The story of how man grew a bigger brain is, perhaps ironically, mirrored most closely by that of the parrot.

Birds and animals generally either keep their offspring in the womb or in an egg until they are well developed, so that the newborn are already semi-independent, or else they bring helpless infants into the world and put a considerable time into feeding and protecting them as they grow up. Which of these two strategies a species adopts depends very much on its life-style. Ducks, for instance, nest in exposed places but live by waters rich in food, so ducklings are born with the ability to waddle immediately into the nearest pond. Likewise, horses live on the hoof on exposed plains, so give birth to foals that can stand and see within hours. On the other hand, the offspring of rabbits, sparrows, and cats, are born blind and helpless. The young are hidden away and fattened up for several months until they can care for themselves. Depending on which of these two strategies a species follows, the brains of its young tend to do most of their growing either before or after birth.

Parrots are an exception in that they lay eggs rich in yolk so that the developing chicks can build up big brains while in the egg, but the hatchlings are still born helpless and stay in the nest being fed by their parents. In this way, parrots double up on brain growth to become, along with crows, the smartest of birds—although they pay the price of such a K strategy by raising fewer chicks than is usual for birds.

As a class, apes have followed the pattern of horses and ducks rather than animals that give birth to physically helpless offspring. The ape brain develops rapidly in the womb and grows only another 30 or 40 percent after birth. Probably this pattern was inherited from the primate ancestors of the apes who would have needed well-developed offspring capable of immediately clinging on to their mothers as they scampered through the treetops. But, like the parrots, this early start to brain development gave man a chance to double up on brain growth. It did not take much for evolution to reset the growth clock and keep the brain ballooning long after birth.

Even while still in the womb, a human brain grows faster than an ape's. At birth, the human baby's brain weighs about 14 ounces, which is already bigger than an adult ape's. This is probably as much growth as is possible while in the womb, for a human baby's head is dangerously large to slip easily through its mother's pelvic girdle, showing how fine evolution has already cut human brain development. The newborn baby's rather soft skull, with an unfused gap down the center between the two halves, is yet another hazard, but one allowing the brain to keep on expanding before the bones of the skull become set. During a human baby's first year of life, the brain keeps growing at the fetal rate and quickly doubles in size. By the time a child's brain has reached almost full adult proportions at the age of six or seven, it is more than three times the birth weight. Remember that an ape brain grows by only about a third after birth.

So far we have been stressing the increase in the number of brain cells that comes with the resetting of growth clocks. But another very important way in which timing changes make the human brain special is the delay in developing the myelin sheaths that surround and insulate most nerve pathways. Myelin is a fatty protein that is wrapped around nerve branches like plastic around a copper wire. This sheathing stops an electrical pulse from leaking out and triggering other nerves before reaching its proper destination. A thick coating also greatly speeds up a nerve impulse. Some long pathways, such as the nerves from our spine to the muscles of our hands, can kick pulses along at over 120 miles per hour. Many baby animals are born with a lot of this myelin insulation still unformed and myelinization can take days, weeks, and even years to complete. This delay enables the animal to be flexible in the way it learns to see, taste, hear—and even think—about the world into which it is born.

During the time in the womb, animal brains sprout a tangled web of neurons. Each neuron reaches out with hundreds and even thousands of fine branches called dendrites to wire itself up to fellow brain cells. In the days and weeks immediately after birth, this dense forest of almost random connections is ruthlessly hacked back to form neat pathways in response to the world the infant finds itself in.

Scientists have done experiments in which kittens have been brought up in a room where everything is painted with vertical stripes. When presented with horizontal stripes a few weeks later, the kittens simply cannot see them; their eyes have become wired to be sensitive to vertical stripes. So when the kittens reach the horizontal edge of a chair, they often walk straight over, blind to the drop. And because myelinization has fixed the distorted pattern of nerve connections in place, the kittens can never adapt fully to the real world. The fatty sheaths have set the seal on a sense of vision that expects the world to have only upright edges.

This fine tuning of a newborn brain applies equally to humans. Our sense of sight is so plastic during the first five years of life that bandaging the eye of a two-year-old child for even a week can lead to permanently poor sight in that eye, while bandaging at birth can lead to blindness. With no messages coming from the eye to help sort out the random connections in the visual areas of the brain, the visual pathways stay scrambled.

This need for stimulus to tune the brain to the outside world applies most obviously to vision, but a certain amount of pruning of nerve pathways to adjust to the world goes on in all the sensory pathways of a newborn child. The reason for this sensitivity is that it gives youngsters a chance to build the most suitable brain for the world—and indeed, the body— into which they find themselves born. By being flexible, the nerve pathways do not have to be wired in perfectly from the start—a precision that is anyway beyond genetically controlled growth. Instead, they can "feel" their way into the job. However, once this basic pattern has been set in myelin, it takes very considerable training and effort to change it. It is as if deep grooves have been cut in the brain's processing surfaces to channel the flood of life's experience down familiar pathways.

The time after birth when myelin fixes the pathways in place differs tremendously between species. Some animals, like horses, are born ready to see and run within minutes and so must have most of the insulation work completed at birth. Other animals do not even open their eyes for the first few weeks of life. The timetable for myelinization can also vary around different parts of the brain. The brain zone that deals with touch might be fixed before the visual areas, which in turn may be set long before the motor patches that control the movement of our limbs.

On the whole, human babies delay all myelinization for a lot longer than other animals. A human infant is helpless for many months, and some parts of the brain are exceptionally slow to become myelinized, taking over two years. Significantly, the areas associated with hearing and speech are plastic right through the toddler years. This gives children a unique opportunity to be shaped by their social world: By a simple adjustment to the on/off growth switch of myelinization, evolution has allowed all youngsters to fine-tune their ears and vocal equipment to adult rhythms of speech. Human infants do much of their learning of language at the impressionable age of two, and failure to get their speech equipment wired up correctly at this stage can leave them with permanent difficulties, just like the kittens lost in their vertically striped world.

The rare cases where children have been brought up by animals in the wild show the drastic effect of such conditioning. In India, two little girls were found who had been raised by wolves. They ran around on all fours, howled, would eat only raw meat, and slept curled up all day. Soon after being taken back into human society, both of them sickened and died. More recently in Uganda a six-year-old boy was found who had been brought up by vervet monkeys after being lost in the bush during a civil war. He moved on all fours and chattered like a vervet. After a year in human society, he had learned to wear clothes and haltingly to walk upright, but he still had not picked up a word of speech.

It is true that the brain always keeps a certain degree of flexibility. Stroke victims, for example, are often paralyzed or brain damaged by the effects of blood clots in the brain but with intensive teaching and hard work, some of the old skills can be taught to other parts of the brain. However, this sort of relearning is possible only with the support and skills of the modern world. In nature, myelinization is expected to fix the basic pathways of the brain for life.

Having looked at the mechanisms behind the rapid rise of man's brain, we should complete the story of man's evolution. When we left the hominids, they were camped on the African veldt about two million years ago, busy with the social advances of food sharing, pair bonding, and dividing up the chore of food gathering. Following these subtle social advances came the more familiar landmarks in the human story as early man discovered the technological advances of tools and fire. These new skills completely changed the name of the game yet again.

The use of tools has some history in other parts of the

animal kingdom. Birds, for example, are great users of tools in the broadest sense. Not only can most birds build nests but they can also do things such as flicking snails against a stone to crack open the shells, breaking off cactus thorns to pry insects out of cracks in trees, or tossing twigs on a pond to attract fish. There is, however, something stereotyped rather than thoughtful about these actions. They are instinctive behaviors wired into the brain at birth rather than discovered tricks.

Mammals also have instinctive behaviors. Kittens, for example, make instinctive pouncing and killing movements when they play and human babies have stereotyped sucking and clinging responses that are wired in. But higher mammals are also able to learn. The most humanlike examples of tool use come, not surprisingly, from chimpanzees, whose most sophisticated trick is perhaps the use of long grass stems to fish termites out of underground nests. Termites, a rich source of protein, emerge from their concrete-hard mounds during the rainy season for brief mating flights, when animals like baboons cluster round to grab them. Chimps, however, have learned to reach termites deep inside the nest by breaking off a grass stem of just the right size and poking it into a nest hole. The termites will angrily grab the invader and the chimp can carefully pull out the stem and suck the struggling insects off it.

Chimps use tools in many other ways. They can break open hard-shelled nuts by bashing them with stones, pick their teeth clean with a twig, mop up water from the bole of a tree with a handful of chewed leaves, or thrash around with a stick to scare other chimps. They can use a log as a ladder to get to something just out of reach, flick sand or throw stones to frighten away other animals, and even wipe their bottoms with leaves after a bout of diarrhea.

Some of these skills are discovered by a chimp itself but

most are learned from imitating others. A youngster will watch its mother fishing for termites and then pick up an old grass stem to have a go itself. Once one chimp has thought up a skill, it is likely to be noticed by others and become general currency in a group. When such a skill has become established in a troop, it is probably passed down the generations and might well last for thousands of years.

A well-documented case of how a skill can be started and then spread is that of a monkey troop in the snowy north of Japan. Tourists would throw slices of raw potato on the sandy ground for the monkeys, who had to brush the dirt off the potato before eating it. One female—a monkey "genius"— discovered that by scooping up a handful of potato and sand and dunking it into the nearby sea, she could wash the sand away. Before long, this potato washing had spread to the young monkeys in the troop who had noticed what she was doing; the older monkeys did not pay attention and never learned the trick. Potato washing was even adopted by neighboring troops, probably spread as a result of the common practice of young monkeys swapping groups to avoid problems of inbreeding.

Our closest relatives, the chimps, show they are quite capable of both making and using a tool, such as a termite stick. However, a certain absentmindedness or lack of concentration characterizes their efforts; they do not seem to think ahead and carry tools with them, nor do they try to make more elaborate tools out of the sticks and stones around them. These shortcomings suggest that chimps live very much in the present, lacking the foresight of humans that would allow them to progress to a humanlike use of tools. Nevertheless, the level of tool use by chimps indicates that our hominid ancestors must have had the brains to make at least some use of tools from the time they first walked upright about four million years ago. Hominids also had the considerable advantage over apes of being physically much better equipped for handling tools. Not only did they have free hands to carry tools around with them as they walked across the grassy plains, they also developed an opposable thumb—possibly as an adaptation to plucking grass seeds with the fingers—which gave them a far more delicate grip.

The first tools man might have used are often thought of as stone axes and spears, probably because the best preserved evidence of early tools is in stone, while spears fit the aggressive hunter image that we have long had of primitive man. Yet a band of hunter-gatherers has a greater need for less dramatic implements, such as baskets for carrying food back to camp or slings to carry babies around while trekking across the African plains. A basket might be as simple as the curved strip of bark used by Australian aborigines out collecting beetle grubs; a sling could be a loop of vine or animal skin. Neither is as likely to be preserved as a flint hand ax—and in any case, it would probably not be recognized for what it was—but baskets and slings are more important than stone tools to the everyday life of a hunter-gatherer tribe, and hominids may well have been using them for millions of years before they started chipping away at flints.

Other likely early tools include an antelope's shoulder blade or a sharp stick to dig up juicy roots hidden in the hard earth or to break open termite mounds, crack a turtle shell, or kill a snake. If hominids made it a habit to carry around the right sort of stick and a basket, it would not be long before they were bringing a great variety of food back to the camp. These seemingly crude tools would have made a tremendous difference to the efficiency with which early man could collect food and the number of children he could bring up.

The first solid evidence of the making of stone tools comes about two and a half million years ago with the appearance of *Homo habilis*, the aptly named "handy man." Our knowledge of *Homo habilis* comes from a few skull fragments and teeth, so how he fits into the family tree is debatable, but he appears to have been widespread in at least East Africa. He stood about five feet tall, with a brain of about 21 ounces— about twice as brainy as an ape but still less than half as brainy as modern man.

Habilis used the simplest of stone tools, no more than a hand-sized rock with a chipped edge, but even this would have been a huge advance, enabling him to chop up a large animal carcass or scrape clean an animal hide. This does not mean that *habilis* was necessarily a big-game hunter. He could have wanted the stones to crack open nuts or scavenge the leftovers of an antelope killed by lions. The significant point is that *habilis* was prepared to put some effort into fashioning tools. He was starting to think ahead and plan for future actions. Also, if he had the wit to shape a stone into a blade, he certainly would have been able to use slings, baskets, picks, and other tools. However, *Homo habilis* seems to have got stuck at a certain level of tool use. He did not progress from his simple chipped rocks. Perhaps the climate and his diet did not push him into inventing anything more elaborate. Whatever the reasons, for about a million years, *habilis* appears to have lived quite happily alongside his close cousins, the vegetarian *Australopithecus* family, with little change in brain size or culture.

The next big step forward came about one and a half million years ago with *Homo erectus*. *Erectus* had a larger brain than *habilis*, halfway between *habilis* and modern *Homo sapiens* in size, and was also taller, standing about five feet six inches. Otherwise, *erectus* looked very similar to *habilis* and the physical changes could have been a natural progression in size due to the steadily better diet of early man. What marks out *erectus* is a big step up in the sophistication of the tools he used. Coincidentally, the various *Australopithecus* species rapidly disappeared—whether through hunting and competition from *erectus* or the start of the harsh ice ages is not certain.

The stone tools carved by *erectus* came, like those of *habilis*, in only a limited variety but they were made with a lot more care. The favorite was a teardrop-shaped ax that could be cupped in the palm of the hand. Considerable effort appears to have been made to find the best-quality stone so that the axes would have sharp, hard-wearing edges. Again, the stone axes of *erectus* have been well preserved in the fossil record but were probably not the most important tools that *erectus* used. With his level of intelligence, we can assume he would have used animal bladders to carry water around, made snares to trap small game, and worn skins as clothing.

Erectus often lived in the ready-made shelter of caves but he also set up campsites. A ring of stones found in Tanzania appears to be evidence of hut building over 1.6 million years ago, but the single most important advance made by *Homo erectus*—especially with the sudden arrival of the ice ages— was the discovery of fire.

Learning to control fire seems a mysterious step. It is hard to imagine how the first person came to rub a few twigs together and start a crackling blaze. Almost certainly, for a long time early man relied on naturally started fires. Brush fires are commonplace in many parts of the world. They can be sparked by lightning or even by the heat of rotting vegetation, and once *Homo erectus* came across the smoldering remains of such a brush fire, it would not be too difficult for him to learn to keep a fire going. Like the Congo pygmies and other hunter-gatherers of today, the women could carry an ember from the previous evening's fire wrapped up in green leaves during the day's journey. Then a couple of twigs and a few puffs of breath would be enough to make a fire flare up again when the group was getting ready for the night.

No doubt much ritual and superstition would surround the

keeping of the fire to make sure the precious spark was never allowed to die out. Later, the making of fire could be refined with metal strikers and dry tinder or the twirling of a twig against a block of soft wood. Yet even in modern times, some primitive tribes like the Andaman Islanders cannot light fires and rely on natural brush fires as a source of embers.

Once *Homo erectus* mastered fire, he was catapulted into a new way of life. Fire warmed and protected the tribe at night. Food could be cooked, making tough meat and roots soft enough to digest, which meant not only that *erectus* could eat a wider variety of food but also that both the very young and very old in a group could live on a high-quality diet. Also food could be preserved for the hard times by smoking it as it hung from the rafters of huts. Fire would have had a dramatic effect on general health: Clothes and bodies could be dried and so lessen colds; the smoke would drive away disease-carrying insects like mosquitoes; parasites in meat would be killed by cooking. In dozens of ways, fire must have proved a quantum jump forward for *erectus*.

Before *erectus*, the *Homo* and *Australopithecus* families appear to have been confined to Africa. But *erectus* quickly spread across the world. Fossil finds such as Java man and Peking man in Asia are representatives of this great migration about a million years ago. *Erectus* also spread through the Middle East, the Mediterranean, and Europe. With fire, tools, shelter, and clothes, *erectus* was equipped to leave the tropics and live in colder climates.

Erectus also eventually became a big-game hunter. While we cannot be sure of *habilis*'s hunting habits, archaeological digs in Spain have revealed that about half a million years ago bands of *erectus* would set fire to the grassy plains at the mouth of a valley and stampede animals into the swamps. Once the herds of elephants and antelope were trapped in the mud, the hunters could move in and pick them off. Archaeologists have found the charred remains of the grass and the neat piles of elephant bones left behind after such an organized hunt.

The reign of *Homo erectus* lasted over a million years, from 1.5 million years ago to about 100,000 years ago, during which time *erectus* must have been hardened by battling through the ice ages. The bouts of ice and drought placed a tremendous strain on most mammals and over half the world's mammal families disappeared in a mass wave of extinctions. Yet *erectus* managed to survive and flourish because of the flexibility his intelligence gave him. However, even *erectus* eventually had to give way to a successor.

Homo sapiens or modern humans arose right in the midst of the ice ages and started a second, even more successful sweep across the world. Once again the journey probably started from the cradle of Africa, where a slimmer and even smarter breed of *erectus* evolved into what we now know as modern man. This new race then spread like wildfire across the face of the world, taking over the lands settled earlier by *erectus* and even reaching virgin territory in the Americas and the Southern Hemisphere.

For *Homo sapiens* to replace a successful race like *Homo erectus* so quickly, he must have had a key advantage. Clearly, his extra-large brain—about a third larger than the brain of *erectus*—played an important part, but as we have seen, the evolution of larger brains is tied to the need to gather enough food to fuel all the extra hungry neurons. The brain of *erectus* had proved sufficient for over a million years, so the sudden expansion in capacity with the appearance of modern man must have been linked with a dramatic new ability, which once again changed his life-style. We cannot be sure what the key difference was, but a good candidate for the cause of this next quantum leap would seem to be language. Simply put, the first hominids discovered upright walking and social graces, *Homo habilis* discovered tools, *Homo erectus* discovered fire, and *Homo sapiens* discovered language.

Human speech is commonly recognized as the dividing line between ourselves and the rest of the animal world. It must also be an important dividing line between us and our ancestors. The reason why the ability to speak is such a sharply defined boundary goes deeper than the mere existence of a method of communication. It is what we have done with language that counts. As we shall see, language paved the way for all the special human abilities that we so value— abilities such as self-awareness, higher emotion, and personal memories. Language provided the building material with which evolution could write revolutionary new software for the hardware of the ape brain.

To see how language could have so dramatically remodeled the mind of a hominid, we need first to take a closer look at the various basic mental abilities such as learning, memory, imagination, and thought. We need a good understanding of the foundations of the mind to appreciate the structures that were then built on top of it with language.

TWO

The Fisherman's Net

So far, we have taken an evolutionary perspective on the human mind. Now we need another vantage point from which to catch a clearer view of ourselves. One approach might be to look more closely at the separate components of the mind. Like stripping a car engine down to its pistons, crankcase, and spark plugs, we could split the mind up into what appears to be its various parts: memory, learning, imagination, perception, rational thought, emotion, self-awareness, and so on. However, there is a danger here that we may once again fall into the trap surrounding the use of the word *mind*. That is, once we start labeling what the brain does, we are tempted to treat these abilities as if they were phantom objects with some mysterious existence separate from the brain.

This turning of verbs into nouns is not a problem with everything that the brain does. For instance, we can accept that the brain is sleeping or that it is dead without getting tangled up in a search for phantom objects called sleep and death. We know that sleep and death are general labels to describe particular ways of behaving—or, indeed, not behaving. However, we are too close to our own conscious experience of life always to talk about our mind so clearly. It is as if we were trying to see our face without having a mirror; we can see everyone else's face quite clearly, and so build up a general idea of what faces should look like, but our own is still frustratingly invisible.

Once we start examining the different components of the mind, we find ourselves talking about memory instead of the act of remembering, imagination instead of imagining, self-awareness instead of being self-aware, thoughts instead of thinking, and attention instead of attending. We set out on a dozen new ghost chases trying to run down the elusive mechanisms we feel must be tucked away somewhere in the gray crevices of the brain. But the idea of a whole lot of different bits of the brain doing different jobs is a false one, drawn from images of motor engines, clocks, and computers. The different parts of the mind are just different facets of the job being done by the same underlying flesh and blood organ.

In searching for a new vantage point to look at the mind, we need to keep this underlying unity to the fore and also stress that we are talking about active processes. What is wanted is a fundamental unit of currency, something like an atom of thought or a mental building block, to describe all the mind's various aspects.

Perhaps our basic building block could be a single brain cell (or neuron, to use the technical term), since we already know something about the chemical and electrical workings of individual nerve cells and since neurons are indeed the common building material of the brain. But just as a detailed description of a tree does not help much in describing a forest, so understanding everything about the behavior of a neuron does not equip us to describe what we see when we step back from the neural jungle. We need a new vocabulary with a broader sweep. Rather than a single cell, a better building block is probably a whole mesh of neurons. And because we are talking about active processes, we want to focus not on any old clump of nerves but rather on a network of cells that is actually working together to do something.

Out of the billions of neurons that make up the brain, it is only the nerves active at a particular moment that are of real interest to us. We are seeking the fleeting patterns of neurons that stand out from the gray background as they jangle together in response to a sensation, a thought, a memory, or an action. These fleeting waves of firing across the brain are like a television screen. Look closely at the face of a TV tube and it is easy to see the thousands of dots glowing red, blue, and green that make up the picture. Each dot can take part in many different images—perhaps one second a cow and the next a speedboat. We can know everything there is to know about television-tube dots and yet miss completely the fleeting patterns playing across the screen. Similarly with neurons, we need to concentrate on the fleeting images rather than the individual cells.

To help get a better feeling for how networks of nerves act together to create our conscious experience of life, we can look at how brain cells gang up to tackle the most important of jobs—sight. Vision seems a simple and effortless act. When we look around a room and catch sight of all the familiar furniture, it feels as if all we have to do is open our eyes and let the shapes and colors flood in like summer sunshine through a window. But the polished ease with which we see things is deceptive. Behind the scenes, an army of brain mechanisms has been beavering away to splash this picture of the world across the back of our brains.

The first thing most schoolchildren learn about vision is

that the eye is like a camera, with a system of lenses that focuses an image upside down on a light-sensitive back plate. But once the image arrives at the retina at the back of the eyeball, the similarities end. A camera turns a living three-dimensional world into a flat record on a slice of film. Vision, by contrast, turns the flat image that hits the retina into a living three-dimensional conscious picture spread around the brain. And while the camera may never lie, the eye—or more correctly, the eye and the brain areas behind it—exaggerate wildly. For example, the eye artificially sharpens up the outlines of everything we look at.

This can be demonstrated by looking carefully through one eye at a sharp-edged object against a bright background—such as a building against the skyline or this book against a window. The edge of the book or building turns black while, by contrast, the adjacent sky goes luminously pale. This sharp black-and-white rim of contrast where the two shapes meet can only just be made out, but so strong is the effect that it is there even if a white-covered book is held up against a white sky.

The 125 million light-sensitive nerves of the retina do not simply pipe a TV-screen image of the world back to the conscious brain. As soon as a pattern of light falls on this intelligent surface, millions of nerves start pulling the pattern to pieces. Along an edge where a dark block of color hits a brighter patch, the nerve cells lining the retina work to exaggerate the contrast. The nerves running at a low buzz in the dark shadow of the book or building sense through a maze of inter-nerve connections that they are lying next to nerves firing brightly in response to the sky. This causes them to switch right off and falsely report complete blackness to the higher visual areas. On the other hand, when the sky-detecting nerves sense the black edge alongside, they react by firing even more wildly and send back signals of luminous brightness. This exaggerated firing only happens across the cells straddling the edge. A few nerve cells away from the edge— into the uniform brightness of the sky or the dark of the book—all the cells will be surrounded by like-minded neighbors and so will respond normally.

By this simple mechanism of cross talk between nerve cells, the eye puts an artificially sharp outline around every object we see and so makes the world stand out with an unnaturally clearcut precision in our minds. Instead of giving accurate readings of brightness, as would an inanimate light meter, the nerves of the retina have evolved a special wiring pattern that makes them react with extra sharpness to any line or contour in our field of view.

This little trick of sight is so much a part of vision that we never normally notice it. Moreover, it is only one of the many inventions of sight that have evolved to help us see more clearly—or rather, more intensely—than would naturally be possible. The act of seeing is littered with tricks that add contrast and make straight lines, smooth curves, or sudden movements stand out. And at a higher level of processing—in the brain itself—more tricks give us a binocular sense of depth, blend together the colors we see, and assemble all the blotches and patches of light into recognizable objects. Later we will see how the output of all these different "feature detectors" is used to give us the intact and whole view of the world we experience every time we open our eyes.

Once we look closely at how the brain handles vision, the oddities and exaggerations of the process quickly mount up. A sense that seemed so effortless and accurate turns out to involve hard work by many different brain mechanisms. The main job of these mechanisms seems to be to distort for the sake of emphasis or to cover up any defects in the process— such as the blind spot right in the center of the eye, where all the blood vessels and nerves break through the back of the retina to leave the eyeball. The hole exists but the eye pretends it is not there and this gap in our vision is known to us only from carefully designed experiments where a moving line seems to jump quickly across the small gap.

Our eyes play a further trick on us: Only the very center of our field of view seems to be in perfect focus; everything outside the direct line of our gaze seems blurred or vague. This is not because our eyes are like a camera and the center of our eyeballs is the only place where things are focused properly, but because there is a pinhead-sized pit in the middle of the retina, known as the fovea, which is packed with almost as many light-sensitive cells as the rest of the retina put together. The clarity in the center of our field of vision is therefore a measure of the number of light-receptor cells at work on that particular part of the image falling on our retinas.

These various mechanisms for exaggerating life and papering over the cracks of vision happen in the eyeball. The story becomes even stranger when the images of the world start hitting the brain to form our conscious experience of life. Before this can happen, both retinas have to sweep together their partly digested images of the world and pump them down the thick cables of the optic nerves as a set of coded messages to the brain. Having broken the world down into a patchwork mosaic of edges, shifting shapes, angles, and blocks of color with the 125 million light-sensitive cells lining each retina, the eyes funnel the fragments down the one million fibers of the optic nerves to be reassembled in the brain as our conscious picture of the world outside.

The optic nerves arrive first at a lump of tissue at the center of the brain known as the thalamus. This knot of nerves buried deep inside the brain acts like the gateway to the wrinkled cerebral hemispheres above it and handles messages not only from the optic nerves but from most of the other senses as well. The thalamus fans the information-loaded optic nerves across the correct part of the gray matter—the cortex—for analysis.

The cortex—a name taken from the Latin for "bark"—is a thin rind of tightly packed neurons that covers the outer surface of the cerebral hemispheres. It is here that the work takes place that creates the cool, rational, and conscious mind. Underneath the cortex are the pulpy white insides of the cerebrum, which are basically a mass of wiring connecting the cortex to the rest of the brain. Although only an eighth of an inch thick, the cortex is packed so densely with neurons that a pinhead-sized piece holds about 30,000 cells. Seen through a microscope, the arrangement of neurons into upright columns looks like strands in a thick pile carpet, with each column having many connections to neighboring columns to create the sort of cross-talk networks found in the eye. However, the input is not light but messages rising from the lower parts of the nervous system, for the cortex is not dealing with raw sensation but with images of life that have already been at least partly processed.

If all the deep wrinkles in its surface were ironed out, the cortex would form a sheet big enough to cover a tabletop. This tablecloth-sized processing surface is divided up into different zones for processing the senses. For instance, an inch-wide strip running from ear to ear over the top of the brain deals with skin sensations. Nerves from all parts of the skin arrive at this strip to be neatly fanned out in the form of a map of our body's surface. This map is greatly distorted in a way that matches how some parts of our body are more sensitive to touch than others. For example, our hands take up a large proportion of the cortex map because we have so many touch-sensitive nerves on our palms and fingertips, whereas the back is represented by a smaller area than that for the hands because we have few sensory cells sending in mes-

sages from the skin on our backs. However distorted, the map
is still recognizable, and if the cortex cells could light up like
a neon sign, we would see a misshapen but human figure
stretched over the center of our brains.

Right at the back of the head is the visual area or occipital
lobe, on to which the flow of images coming down the optic
nerve from the eyes is projected. As with the misshapen body
sprawled over the top of our brains by our sense of touch, the
nerves of the visual cortex are triggered into painting a distorted but still recognizable image of the shadows falling on
the back of the eyeballs. This visual map on the back of the
cortex is only the start of the whole act of consciously seeing
things. The brain makes its initial mapping on a postage-
stamp–sized square of cortex and then, like ripples from a
pebble dropped in a pond, this image is repeated in waves
that fan out across the back surface of the brain. Rather than
holding a single clean image, the cortex is awash with echoes
like a picture viewed through a fractured prism.

There is a good reason for this repeated mapping of what
the eyes see. The central postage-stamp–sized zone of the
visual cortex makes only a first stab at analyzing the image.
It is then set upon by encircling bands of nerves which echo
the image and are wired up to pick out specific details. For
example, one of the surrounding zones might be looking for
the clues that give us our three-dimensional sense of depth;
it can tell how near a chair or table is to us from slight differences in focus and angle. Another zone might be tuned to
the job of compensating for the turning of our heads, so that
the world appears to stay fixed as we swivel rapidly around.
Yet other bands make sense of the shapes we see, telling us
which jumbles of edges and patches of color might belong
together to form some object such as the family cat. Each of
a dozen or more bands is set up to do a special job and con-
tribute to the analysis of the image captured by our eyes.

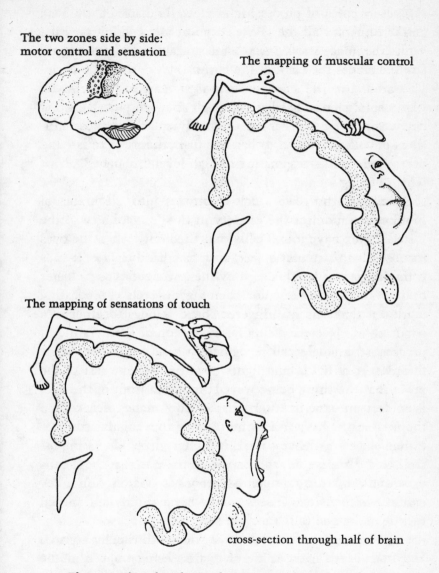

The two zones side by side: motor control and sensation

The mapping of muscular control

The mapping of sensations of touch

cross-section through half of brain

The sense of touch and control over movement as mapped out across our brain surface—note the exaggerated amount of space devoted to mapping "sensitive" parts of the body such as the mouth, feet, fingers, and tongue. Also, the very small region given over to cells reacting to messages from the back and skull.

These ripples of processing start at the back of the head and fan forward all the way to a point just behind our ears. While the image starts out as a recognizable map of the world painted across the back of the cortex, by the time it has rippled around to the side, the map is starting to break down. The mental picture has been through so much processing and condensing that the echoes originally splashed across millions of cells die down to become just clumps of cells—and even single cells—responding to high-level "thoughts" about what is being seen.

Scientists who have stuck electrodes into the brains of monkeys to monitor the activity in this forward area of the visual cortex have found cells that fired only when the eyes focused on a particular object, such as a hand or face. It is as if the image of a hand is seen by the lower zones as a cluster of lines and patches of color, then higher levels of processing work out that this five-fingered shape belongs together as a single solid object, and finally, at the top of the pyramid of processing, a single cell recognizes a hand—from any angle in any sort of light—and lights up to inform the rest of the brain that the thing being waved around in front of the eyes is in fact an experimenter's hand. How many such cells a monkey might have in its brain, tuned to recognize such familiar objects as leaves, fruit, or hungry leopards, is impossible to determine, but the important point is that these cells exist and can bring a pinpoint sharpness to vision. A monkey can be positive when it sees a hand because only one special cell lights up and tells its brain so.

Our visual impressions of the world are thus being twisted and fragmented even as they are first being trapped at the retina. A mosaic of images is then piped to the back of the skull and reassembled in an artificially sharp map. If we lounge back in our chairs and clasp both hands behind our head, our fingers will be a mere half inch of skin and bone away from

Messages from the eye are first fanned out across the visual cortex on the back surface of the brain.

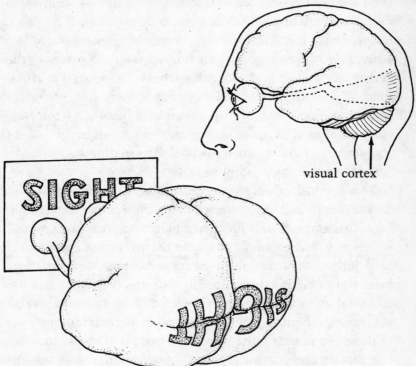

visual cortex

This first mapping of vision is "projected" upside down and back to front on a postage-stamp–sized patch of brain cells.

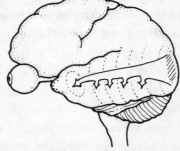

Then this initial projection ripples forward, passing through bands of processing, which extract different details such as color, depth, and shape.

the dark network of brain cells in which the bright conscious
sense of life appears to flicker. If we were then to run a hand
toward our jawbone, we would cross perhaps a dozen more
visual maps where different aspects of sight, such as color,
outline, and fullness of shape, were being emphasized. Finally, if we kept on going, our fingertips would rest over the
cells that "know" and recognize whole objects in our visual
field. It is all these bits and pieces of activity taken together
that create the sensation of seeing. Somehow, all the fragments are fused to make up our own conscious experience of
the world around us, so what is a shattered mosaic of brain
activity appears as a seamless whole.

The way vision works is typical of how the brain reacts to
all sensation, and since we set out with the purpose of finding a fundamental unit for capturing the way the brain works,
we seem to have found our focus: the straggling networks of
cells jerked alive by each passing sensation. Now what we
need is a name for these fleeting patterns. Psychologists use
the word *engram*—from the Greek for "trace"—and also talk
of chunks, schemata, scripts, and processing frames, but none
of these terms feels quite right. The word that seems to strike
the correct note is, simply, *net*, because a net is an apt description of the tangled mesh of neurons that makes up our
mental building block and it can be any size or shape, from
a web of half a dozen neurons to a vast network of millions
of firing cells. Furthermore, nets are used for trapping and
the main job of neural networks is to trap information.

A net lights up either as a map of the outside world or as
a web of nerves that traces out a memory. It is the dancing
webs of nerve activity that power the mind, so if we can
focus on the underlying behavior of these nets, we should get
a feel for the way the mind works, and eventually come to
see the different aspects of the mind in terms of nets. For
instance, we could talk about learning as being the setting

up of a net, memory as the reactivating of a net, attention as the act of focusing on a particular net, and sensation as the fleeting flicker of a net. Thought could be the logical linking together of several nets and self-awareness the forming of a memory net about the self.

Looking in more detail at the "natural" behavior of nets, we have already seen that they are self-defining. The straggling network of cells that comes to life for a particular task is what defines that net. It is created the instant it lights up and vanishes as soon as it fades into the gray background. Nerves tire and fade even when stimulation continues, so it is quite normal for all nets to be short-lived in the brain. Also the same neuron can be part of many different nets, so—like the dots on a TV screen—a neuron in the visual cortex can be a dark patch on a cow's left ear one minute and the corner of a speedboat the next. Although the neuron can of course take part in only one active net at a time, it might switch between nets in rapid succession and a well-used processing surface, like the visual cortex, dances to the tune of many different nets each second.

Another important property of nets is that they have no true boundaries. A net—much like vision—has a sharp central focus of nerve activity that blurs into quietness at the edges. There is no precise boundary; the rustling of nerves at the fringes fades into the gray background. It is like a pebble dropped in a still pond, sending out ripples that spread and ebb away. If we agitate the central focus of a net, it is like dropping a larger pebble. We will increase the total spread of the net as the stimulation excites more of the nerves out at the fringes. When we prod a patch of neurons, the patch will spread like a stain. Or if we have a long straggling network, with nerve fibers stretching right across the brain, agitation at the center will send jangling tendrils even deeper into the far corners of the skull.

A further important point about nets is that they are not fixed to the brain surface. Just as images on a TV screen can shift while all the flickering dots that create the apparently moving picture stay in the same place, nets also "move" in this way. The central focus of a net can appear to glide across the brain's surface as the clump of cells that originally became active quiets down and a fringe group of cells raises its tempo to pick up the beat. The brain surface would stay put but the active heart of the net would appear to spread or flow across the gray matter.

This brings us to perhaps the most crucial aspect of a net. We have said that a net is always defined by the web of cells active at any one time, but a net is still a blend of both present and past. The path taken by any new sensation depends not just on the shape of the stimulus but also on channels that might have been etched into the brain by experience. Imagine the brain's gray matter when we are born as a vast, smooth wax tablet. As soon as we open our eyes and start to react, the sensations cut grooves in the wax, like rivulets of boiling water. Each new sensation acts like another trickle of hot water, cutting its own track across the tablet. Soon the wax will be laced with channels and new trickles will start to flow into existing channels, making them deeper and broader. Eventually, most of the water will run down these deeply etched grooves and only a very large or very hot jet of water will have the strength to carve out its own track. The brain works in the same way. Experience etches deep patterns in the gray matter. What we have experienced before will flow easily down ready-made channels, while new sensations will find the going hard or tend to be diverted into existing channels.

The brain is not, of course, covered with microscopic grooves like a vinyl record. However, one of the important properties of a clump of nerves is that the cells can become chemically primed to fire more readily if they have fired to-

gether before. Chemical changes take place at the nerve junctions between cells, putting them on a "hair-trigger" readiness
to fire again in unison at the slightest touch. Nerves also
grow new branches to their active neighbors to strengthen
the bonds already formed. Because of these physical changes,
a frequently aroused pattern of nerves is far easier to awaken
than a totally fresh arrangement of cells. As the brain goes
through life, millions of well-trodden paths are worn among
the jungle of neurons. They lie hidden in the gray background when there is nothing to stir them into action, but
these dormant nets live on a hair trigger and will fire into
life again at the slightest prodding.

These hidden patterns give extra body and shape to freshly
created nets, filling in any missing gaps. For example, if we
look at a cow in an open field, our visual areas faithfully map
all the shapes and colors that make up the cow. But if the
cow is standing half hidden behind a fence with a tree cast-

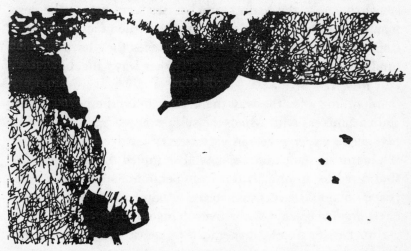

*At first sight this is an unintelligible jumble of lines, but when told
what the picture represents, most people find it clicks into place
with almost photographic reality. The brain supplies the missing
information to flesh out the raw image. (See page 66 for the answer.)*

ing confusing shadows over its back, the net created by mapping this confused image from the eyes might be only vaguely
cowlike. The image fed through to the brain might have only
20 or 30 percent of "cow content." This is where experience
could step in. We have seen so many cows before that the
typical pattern of nerve firing is etched deeply into the brain.
If the rough image from the eye overlaps part of this underlying etched pattern, then the whole pattern could be triggered back into life. In effect, the eye might be supplying a
third of the cow and the memory pattern providing the other
two thirds. Consciously, we would be sure that we were seeing
a whole cow. Its legs might be blocked from view, but we
would know they were there and could imagine what they
would look like; similarly its back might appear indistinct
but our memory could supply the missing details needed to
complete our visual experience of seeing a cow.

This example gives an idea of how nets work. They are
splashed across the brain by the senses, carving out the permanent pathways that we call memory, then fading quickly
from conscious view. But later, the traces they have left behind can be used to flesh out or in other ways affect the nets
that follow.

Importantly for the way the brain works, the permanency
and faithfulness with which the surface of the brain is marked
by fleeting experience can vary tremendously. Some parts of
the brain's surface may be so well trampled that it is hard to
find any underlying pattern left behind. For instance, the
postage-stamp—sized visual cortex is bombarded with nets all
day long, every day of our lives. This constant flow washes
the surface clean of all memories so that the visual cortex
simply dances to the flickering images coming in from the
eyes. But we have also seen how the initial image on the
visual cortex is then rippled forward through increasingly abstract zones of processing, so that the frantic buzz of activity
in the primary visual area is left behind, until there is just

the occasional popping of high-level nerves as they are triggered by the sight of whole recognizable objects such as a cow or a hand. Forward in the quieter zones, each fleeting pattern must be able to leave behind a more lasting imprint on webs of cells.

If we look at the process as a whole, the visual cortex is alive with freshly created shapes and outlines. The ripples from the initial image run forward into zones where the emphasis changes from the display of immediate sensation to the storing of old memories. The ripples are filtered by the deeply etched memory surface and start falling into patterns that become quite specific to, say, cow-shaped or hand-shaped objects. It is as if each net is a pyramid of processing. Each sensation has a broad base rooted in the present with a wealth of detail splashed across the visual cortex. This initial burst of nervous activity is then channeled forward, narrowing down to specific memories and ideas about the world as it ripples across the wrinkled surface of the brain. At its peak, the net touches on the buried world of memories and triggers an echo from the thousands of other cow-shaped or hand-shaped patterns that have been channeled forward to this same point of the brain's surface. In this way, every net can simultaneously have its foundations in the present and a peak of processing that reaches deep into the world of buried memories. We see things as they are but also touch on the many echoes of similar things we have experienced before.

A net, then, has many special properties. It is self-defining and generally short-lived. It has no sharp boundaries and it can spread and flow. It is a mixture of past and present patterns of firing—the present tending to form the broad base of detail; the past, a sharp point of precise memories.

This description of nets was intended to help strip the brain's working surfaces down to basics so that we can see how the nature of nets gives shape to the abilities built on top of them.

We will see how learning, memory, imagination, and the rest
of the mind's facets are simply a reflection of net behavior.
Before examining each of these facets, however, we should
note how much goes on in our minds even when we are doing
the simplest things.

Imagine driving fast down a country road on a warm summer afternoon. Glancing into a field, you are surprised to
glimpse a group of people sitting around in a circle. A second
after flashing past the scene, you are wondering what people
sitting like that would be doing and you might rapidly sift
through, and dismiss as unlikely, a few possible explanations
such as that they were involved in a religious ritual or a scientific experiment. Suddenly it hits you that they were having a picnic. Thinking back, you can almost remember that
they appeared to be eating and you have a strong feeling that
they were sitting on a blanket. And was there not a picnic
hamper in the middle? You would not want to swear to that
but it would not surprise you.

A little later on the journey, you round a sharp corner and
see the top of another vehicle sticking out above the edge of
the ditch by the side of the road. Instantly you get the pan-
icky feeling that a car has crashed off the road, but a split second later you realize that the car is a farmer's tractor. You think

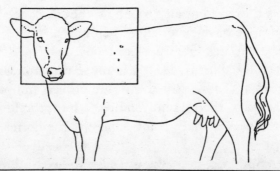

The picture is a cow with head tilted toward the observer.

the tractor has been left in an odd place but is unlikely to have crashed, so you relax back into your seat and speed on again.

Such feelings hit us dozens of times each day. We initially get a wrong or puzzling picture of something we come across and then the correct answer leaps out to show the scene in the correct perspective. It happens so naturally that we rarely take any notice. But what is really going on when everything suddenly falls neatly into place?

It is as if our heads were inhabited by a librarian who sees a woolly picture coming in from the eyes and cannot make sense of it until he pulls down the right book, the one that closely matches what we are seeing. He might flick through books on weird religions and open-air experiments before grabbing hold of the book on picnics, when with a cry of eureka he will tell us that we have seen a picnic. The inner librarian could then go on to tell us what we should expect to find if we turned the car around for a closer look: perhaps paper plates, ham sandwiches, melting butter, and ants crawling up bare legs. We would be armed with a huge catalog of things we would expect to find—as well as things we would not, like prayer mats or oscilloscopes. Equipped with this sort of knowledge, we could easily convince ourselves that we had been sharp enough to spot hampers and blankets even though we had caught only a vague blur. Like using our cow knowledge to flesh out a half-obscured image of a cow, we could flesh out our blurred impressions of the scene with our general knowledge about picnics.

The same sequence of mental events would have happened with the road accident. The first book opened seemed to be the right one but as we got closer, the shape of the crashed car became less and less like what we were expecting until suddenly the librarian had to leap up and hurriedly get down the book on farm tractors. That book told us that the big rear wheels and glass driver's box were definitely tractorlike, although the way it was left down in the gully still needed some explaining.

In the examples just described, the events took place in split seconds. A jumble of sense impressions arrived in the brain and was rapidly matched to the correct clump of picnic memories or tractor memories held in our memory banks. Once the roadside scenes had been put into the right frame of reference, they clicked into focus and we felt we understood what we saw and had opened the door on an illuminating warehouse of knowledge. Although an inner librarian was pictured as doing all the work, this image is a little too active for the brain, which does not have moving parts but the shifting patterns of nerve activity that we have called nets. We need now to see perception and memory in terms of these nets.

The process of perception is probably the best place to start. "Perception" sounds like a fancy way of talking about sensing, but it is a useful term because it makes the distinction that we are never in direct contact with the real world. Everything we see, hear, touch, taste, or smell has been filtered and distorted by the pathways leading to our conscious experience of life; yet, as the description of vision has shown, despite the fragmentary processing, we feel that we see a whole and intact world, or rather, we gloss over the cracks and believe that we are seeing a 20/20 view of life.

This gluing together to create an apparently complete vision is the key to perception. Remember that our visual pathways had to be built up over millions of years, starting from the humble beginnings of a worm's light-sensitive eyespot. To arrive at conscious sight, evolution had to tack new bits on to the existing foundations, adding extra filters and higher levels of brain processing until it got the right mixture of processing mechanisms to do the job. This haphazard approach succeeded because every new filter had actively to fit in with all the rest of the filters. There was no room for individualism. To hold together such a jumble, the filters had to work as a team, with the result that no matter how distorted or fragmentary our processing of vision, the filters force on us a unified view of the world. Trying to isolate the separate filters when we look around is like trying to block the water flow by stuffing a finger up the tap. The filters generate such pressure to produce a unified stream of conscious experience that it is a struggle to break vision down into its components again.

The unifying pressure of the filters can, however, be seen in such visual illusions as the railroad-track picture where two lines are drawn so that they apparently narrow into the distance; if a pair of identical-length bars are then drawn, they will appear smaller or larger depending on whether they are "close by," at the wide base of the lines, or "far away," where the lines converge. The zone of visual cortex that specializes in judging distances is trained to assume that narrowing lines are disappearing into the distance, so it makes unshakable judgments about the size of any object that appears nearby. The illusion is so strong that even if we measure the bars to demonstrate they are the same, it seems impossible to block out the feeling that they are different.

There are illusions for every aspect of vision, from the cube that one minute looks inside out and the next looks outside in, to the picture that is one minute a vase and the next two faces, and the Op Art spirals that seem to spin on the page. Some illusions can be demonstrated only with elaborate setups by psychologists. A roomful of experimental gear was, for example, needed to reveal how our visual pathways compensate for the bobbing of our heads and flickering of our eyes as we move. Without processing areas wired into our wrinkled cortex to do this, the world would appear to lurch alarmingly every time we took a step.

Illusions are not limited to vision but crop up with all the senses. A familiar example occurs when one hand is dipped in ice and the other in hot water, and then both hands are plunged into the same lukewarm water: The iced hand feels as if it is burning, while the hot hand feels freezing. All these illusions seem like mistakes, as if our perfect inner view of the world is being led astray by a breakdown in our senses. But really we are catching out our senses with artificial traps. Our senses have learned how best to process the real world and throw together the unified impression that results in our minds, so if we deliberately put our senses in front of visual traps, we should not be surprised that they try to keep on dealing with the world as usual. While we may tell ourselves that the illusions are leading us astray, we cannot help but continue to feel that, for example, one hand really does feel warmer or the bottom bar on the railroad track does look bigger. Our consciousness is not separate from these sensory mechanisms, like an inner observer sitting in front of a mental TV set. Consciousness is our label for what we experience when these mechanisms go into action and whip up their flickering networks of nerve activity.

Just how much our senses are educated to tackle the world in a certain way—and also how, to some extent, they can be reeducated—is shown by an experiment in which a scientist fitted himself with a special pair of glasses with mirrors that turned his view of the world upside down and back to front. Initially he could barely move without lurching into something, but after a few days he adjusted to this new way of looking and gradually the reversed world came to look "right way up." The scientist could walk around without problems. When he took the glasses off at the end of the experiment, he had to go through the relearning process all over again, for although the brain switched more quickly to the familiar orientation to the world, the visual pathways still took some time relearning what "normal" vision was.

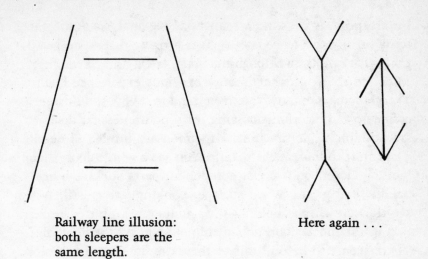

Railway line illusion:
both sleepers are the
same length.

Here again . . .

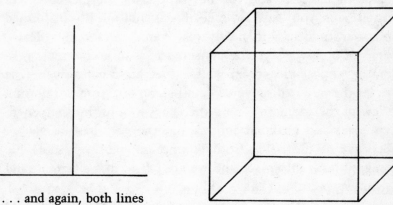

. . . and again, both lines
are the same length.

The Necker cube: the eye switches
between two three-dimensional
perspectives.

Perception is thus more than a passive analysis of the patterns hitting the sense organs. It is a rising tide of nerve activity that swells up through the brain, being filtered and fused to create the froth that is our conscious experience of life. Try glancing around a room at random. You should see a complete and colorful panorama, with no impression that you are building a picture from a fragmentary jumble of detail. Then start to look at the details. You see a whole door; then, looking closer, you see its component parts such as hinges, handle, and grainy wood surface. Looking harder still, you see the screwheads holding the handle on, a blurred reflection in the metal doorknob, a chip of paint off the door edge. There seems to be no limit to the extra detail you can pick out by looking at that corner of the room. Yet drop back to the general picture after you have studied every bump and crack of the door and you have the same simple, complete impression you had in the beginning and not a transformed picture studded with newly discovered microscopic detail. Perception always pushes consciousness up to the most general level possible, so our well-trained brain pathways refuse to let the newly discovered detail stand out from our normal view of the room any more than they would the sharpened-up edges and artificial contrasts at the very base of vision. Because of the rising froth of impressions, every detail is dragged back into place and we are left with a clean-cut and seamless picture of the world outside.

In terms of the idea of nets, perception is the process that creates the nets that make up the flickering sensations of consciousness. The sights, sounds, and tastes of the world batter up against the sense organs, and the eyes, ears, and skin sieve patterns from this sensory assault. Then, in the brain, the numerous fragmented images are put through dozens of different filters so that the brain processes the full richness of life. A rising froth of these nets then creates our inner world of conscious impressions, our conscious plane.

These dancing webs form the foundation of awareness, but as we have seen with the country car trip, raw perception is only half the story. The fleeting roadside view may have been filtered a dozen different ways in the visual cortex so that we could savor all the shapes, colors, and movements present in the scene, but if our conscious awareness had been left at that, we would have had no understanding of what we had seen. It would have been a meaningless jumble of impressions, as if we were watching the shifting patterns of a kaleidoscope or listening with an unfamiliar ear to a classical concerto. To raw perception we need to add the right frame of reference so that we know what it is that we are sensing. We need to unlock the doors of memory and mobilize our knowledge about picnics or car crashes to make sense of the impressions.

This matching of perception to knowledge is usually so automatic that it passes unnoticed. It is only when the process takes time or we get it wrong—such as mistaking a picnic for a religious ceremony or open-air experiment—that we pay it any attention. Our waking lives are spent in an unbroken flow of bridges between sensation and memory. Picnics and car crashes may have stood out on our country drive but our minds would have been a steady stream of thoughts between these two events. A flash across the road may have been recognized as a sparrow winging dangerously low in front of our speeding car. A sudden warmth on our arm may have been recognized as a burst of sunshine as we rounded a corner into the sun. Such observations draw automatic answering echoes in our minds. Links would have been made between conscious sensation and our locked-away knowledge about sparrows or hot sunshine.

Because birds and sunshine are so commonplace, our thoughts probably strayed no further. Nevertheless, once the link was made, our minds could have followed up with a more intense searching of the memory bank. This linking of

impressions to knowledge is at the very heart of consciousness. Awareness is a constant recognizing of every little thing
that our senses place before us. We feel we are conscious not
only because our brains are full of raw sensation but also
because we recognize and understand what we see.

Recognition is perhaps the unsung ability of the mind. It
goes about its work so smoothly and quietly that it is usually
overlooked. However, when recognition is tested in the laboratory, the results can be surprising. Memory and perception are full of illusions, errors, and limits, but humans show
extraordinary powers of recognition. In one experiment, people were placed in front of a slide screen on which about
twenty-five hundred different photographs were flashed for
just a few seconds each. The pictures, ranging from street
scenes to close-ups of dandelion flowers, were then mixed in
with another batch of slides and the experimenter's subjects
had to pick out the photographs they recognized from the
first screening. Despite having seen thousands of pictures in
rapid succession, most people were more than 90 percent accurate in spotting the familiar slides.

People did not show such a high success rate when they
were shown twenty-five hundred photographs and then asked
to remember as many as possible. With nothing to jog their
memories, most could recall and describe only a few dozen
pictures. Yet the recognition figures showed that the information about the thousands of slides must have been stored
away in their heads somewhere. Such results make it clear
that recall and recognition are two different things. Both
are ways of tapping the vast pool of memory that forms
in our heads, but recognition appears to be a far more powerful way of getting at this information. Before looking at
recall in more detail, we shall first see where recognition gets
its power from.

To understand recognition, we should remember how we

managed to make out the half-obscured shape of a cow standing in the shadows of a tree behind a fence. A vague and incomplete image of the cow was splashed across our visual cortex, but though the general image may have been poor, some details were sharply cowlike—the ears and line of the head, for instance. These details would have been processed by the visual pathways in a very particular way. Only a cow's head has the shape of a cow's head, even when seen in a dozen different lights and on a dozen different cows in a herd, so as the head image funneled forward to more abstract zones of processing, it would start to be channeled along certain paths that only cow-head–shaped images would go down. Eventually the net might narrow to a single cell or clump of cells that had learned to respond only to the shape of a cow's head.

The same pyramid of processing happens with a photograph of a dandelion, as in the experiment described above. Dandelion-shaped images are processed by the brain in the same way each time, so whenever a slide of a dandelion is seen, the image is splashed across millions of cells in the visual cortex and from this broad base is sent surging up the same old processing channels. It reaches its sharp peak at the point where all previous dandelion images ended up and delivers a powerful prod to any memories stored there. For the brain to recognize the picture, all it has to do is say: "Aha, these sorts of messages have passed this way before!" As if a light bulb is flashing on above our heads, we will get a powerful jolt of recognition.

The reason why we are so good at recognizing thousands of slides is because our memories are not just jogged by the faint murmuring of a few cells but receive a thumping kick from the upward thrust of the whole visual pathway. By contrast, when we consciously scour our memory banks for recollections of some of the thousands of slides we have been shown, we are relatively weakly equipped for the search. It is as if our memory banks were a darkened room full of objects with us blundering around with a penlight, hoping to bump into what we are looking for. However, when the original experience is repeated and comes crashing up the sensory pathways, it is as if all the lights in the room have suddenly been turned on, so the original memory is extremely easy to find.

In terms of networks of cells, recognition is simply the rippling upward of a net until the net narrows to a sharp point buried in memory. This perception net strikes a net of memory and awakens all our locked-away knowledge about picnics, cows, dandelions, or whatever. The twinge of emotion we experience when this happens—the "aha" feeling—is worth a closer look. How is it that we can trust this feeling as much as we do? Why do we feel certain that we know something, are sure for example that the capital of Russia is Moscow? If we look closely at precisely what happens whenever we feel we have struck on a correct answer, we get the excited "aha" feeling. It is almost a physical sensation, and if pressed, we might explain we just had a gut feeling or a hunch about its being right. We could call this the "recognition" feeling or the shock of recognition. At the instant when an incoming perception clicks with a stored memory, there is a sudden burst of the "aha" feeling, as if all the nerves involved have clapped their hands in happy surprise. The feeling is one of making the right connections.

The "aha" feeling is clearly genuine and can range in strength from a minor twinge to a massive jolt. Normally, spotting a familiar dandelion picture among thousands of slides will give us only a faint buzz of recognition but the shock of recognition would hit us more strongly if we were desperate to find the picture again. If, for example, our lives depended on correctly identifying the slide, we would shake with relief at realizing we had the answer. However, no matter how faint the click of recognition may be, without such a spark we would have nothing to base memory on; one answer would appear as good as another. This means that if the mechanism of our recognition let us down, our conscious minds would be completely helpless to correct the mistake. Fortunately, the workings of recognition evolved over millions of years, and along with most of the rest of the animal kingdom, we would not have survived so long unless the "aha" feeling was a dependable foundation for "knowing" when we know something.

The only common way that recognition does let us down is the puzzling experience of déjà vu—the sensation that everything we are experiencing has happened to us before. With déjà vu, for five to ten seconds everything appears familiar, but we cannot pinpoint the original memories that we are half recognizing. This is because such memories do not exist. Déjà vu is the recognition feeling stuck like a needle in a groove and telling the mind that everything coming in through the senses is familiar. Perhaps the needle gets stuck because something quite close to an earlier experience has struck a half echo in our minds—but there is not a perfect match. The brain keeps struggling unsuccessfully to make the match and the unfocused jangling of our recognition alarm bells colors all the experiences that follow. For the few seconds until the fake feeling of recognition fades, everything that passes through our mind seems to have happened before. The feeling is eerie while it lasts and would obviously cause us terrible confusion if it continued.

The recognition feeling probably lets us down far more frequently in the cozy world of our own thoughts. When we want to solve a problem, we often fill our minds with as much detail about it as we can—like the broad base of detail that is the foundation of a sensory impression. Then we stir it around until—with a flash of insight—the answer pops up. Somewhere deep in memory, the prodding of this broad base of detail has flushed out the right image or fact to solve the problem. We see with a jolt of recognition that the simple answer matches precisely the wealth of detail. In this way, the "aha" feeling is part of all our thinking, which makes it important to ask just how accurate and reliable the "aha" feeling is. After all, it is not uncommon to have blinding insights about life after a few drinks, only to realize later how silly those ideas had been. A happy feeling of recognition may have made the insight convincing at the time but did not automatically make it true.

Also, it is easy to talk ourselves into thinking something is familiar when it suits us. The example of the fleeting glimpse of a picnic shows that our recognition of the scene might have been based more on probability than any real evidence. Our visual pathways trapped only a fragmentary image of a group of people clustered in a field. After mulling the sight over, it seemed more likely to be a common event like a picnic than something strange like an open-air experiment or religious ceremony. This tentative feeling encouraged us to start looking for picnic equipment like blankets and hampers among what we could remember from our brief glance. Quickly we might accept that a faint trace of something in the center of the group was a hamper and so make the whole picnic guess appear more certain. The "aha" feeling of recognition would creep up on us, though if we were honest with ourselves, we would admit we hardly knew what we were looking at. But imagine we had a passenger in the backseat who asked what it was we had just driven by. It would be human nature to sweep all lingering doubts under the carpet and reply with confidence that it was of course a picnic. Luckily, however, the sense of recognition is a mechanism that has stood the test of time, so mostly it is robust enough to withstand the wishful thinking of humans.

* * *

To sum up our story of nets so far, we have seen how the world batters against the sense organs like a stormy sea. The sense organs sieve pattern and detail from this flood and channel it up to the cortex, where the world is re-created in the dancing webs of nerve activity that make up our conscious impressions of life. All of this nervous activity takes place in a split instant. A net travels through the brain so fast that it appears to exist as a whole glittering network of cells with its broad base rooted in the present and its sharp peak plunging into memory. Once the net strikes the right part of our memory banks, the alarm bells of recognition go off and the gates to our warehouse of knowledge swing open.

When we talk about the door of memory being opened, it does not mean that we are suddenly conscious of every bit of stored knowledge. Rather, the sight of a picnic or a cow seems to take us to the center of where all our picnic- and cow-related memories are kept, so that we can start digging around and bring particular memories to light.

What then are memories? Despite likening them to inner libraries, they are more like living patterns etched on the brain by experience, which can be brought back to life with a sharp prod. Memory traces are created when webs of nerves become bound together after responding in concert to a wave of sensation. A net washes up to a peak and leaves behind a residue of connections, like driftwood cast up on a beach. The original sensation rapidly fades, like a wave draining back down the beach, but the flotsam of a memory pattern is left behind at the high-water mark. Afterward, the stimulation of this knot of cells will reawaken the pattern of firing that has been stamped into them. If the sparking of a buried memory net causes the exact same arrangement of high-level neurons to come to life again, a repeat of the original sensation will once more become part of the general clamor of our conscious experience.

Our memories are a huge collection of these buried patterns waiting to be breathed back into life. Millions upon millions of waves of sensation have washed over our minds during our lifetime and left behind their mark. Any time that we reawaken one of these etched traces, we cause the original sensation to be repeated in our heads as if we had put an old record on a turntable.

We cannot, of course, stir all our life's buried memories into action at the same moment. For a start, some key neurons may be involved in hundreds of different memory patterns, and like dots on the TV screen, the neurons can take part in many different images—but only one at a time. Also, as we shall see, there are evolutionary reasons why consciousness is geared to dealing with one memory at a time. This limitation means in practice that for us to explore our hidden store of memories, we have to walk around the darkened warehouse with a match and light up each individual memory as we go, until we find exactly the one we want.

To picture how this awakening of memory works, imagine memory as a vast fishing net which has snared all the fleeting patterns of life's experiences. This vast memory surface is highly organized. Each wave of sensation will tend to reach its peak at a particular spot, depending on which paths the brain's filters sent it down. For instance, the residue patterns left behind from seeing dandelions are likely to be found near similar-shaped memories for other flowers; picnics will tend to be classed with other memories of meals or family outings. Like a library, the memory surface will be sorted out by topic, with closely associated topics clustered near each other, but there is no need for an inner librarian to arrange the shelves. As we have seen, memories find their own way to the correct spot because all similar-shaped sensations are diverted down the same paths.

Imagine this memory net stretched out just below the surface of the sea. Our consciousness drifts above it like a fisherman in a boat, and to bring a particular memory to light, we have to plunge our hands into the water and drag up a heavy handful of the net. As we pull the mesh close to the surface, we can see the watery wriggling of the fish trapped in the net—our memories. All at once, we have lifted the net clear of the water and can see our catch clearly in the sunlight. If we are hauling up the bit of the net studded with picnic memories, we might have pulled to the surface some memories about jam sandwiches or blankets.

The problem for the mind's fisherman is that he has the strength to hold only a small handful of net above the water at one time. To search the rest of the net for other facts he has to let each present handful slip back into the murky depths. Tantalizingly, though we know we are floating above a vast net studded with every sort of fact about picnics, we can see clearly only the bit of memory we have lifted into consciousness at that moment. However, although we know that we have a vast pool of memories of which only a few can be examined in the full glare of consciousness at any one instant, it is not generally a problem since we can always go back to dredge up old memories again.

Another saving grace of our memories is that dragging up a particular memory brings other, closely related memories nearer the surface. For example, by pulling sandwich memories to the surface, our mental fisherman would be able to see the struggling outline of other facts naturally linked with the picnic memories, so a thought about sandwiches might be the start of a chain of thoughts about sticky jam fillings, swarming ants, the slapping of legs, and spoiled afternoons. Potentially, there is no limit to how far the mental fisherman can follow a chain of associated memories across the hidden memory surface. Handful by handful, we can pull ourselves a long way from our original thoughts about picnics to find ourselves examining memories on, say, rocket launchers or coal mining. But to move on, we always have to let our present handful of facts slip back below the water. We can remember a lot of things in life but can hold only a very few of them fixed in our awareness at any one moment.

We tend to overlook how difficult it is to keep more than a few memories alive in our consciousness at any one time. Images tumble through our mind so rapidly, and we can fish out memories with such ease, that this limited capacity does not seem a handicap. It is like a movie film made up of numerous individual frames, each packed with detail, but spun through the projector so fast that the illusion is created of rapid nonstop action. In the same way, if we could slow our mind down, we would see the huge number of painstaking steps that have to be followed to create the smooth illusion of consciousness.

We have tried to capture this rapid mental action by using nets as the fundamental unit of brain activity. The idea of a fleeting net should help us focus on what is happening at a precise moment while also bearing in mind how the stamp of previous experience will be influencing the current firing of a network of cells. But nets capture only the way a single impression or memory trace is handled by the brain. We have seen that the mind is made up of a tumbling cascade of such nets. Like a fast-cutting movie director, we can sweep our gaze around a room and whip up a dozen different nets a second. Every new angle will throw a fresh image on the visual cortex and create a fresh pyramid of processing. We need a new term to capture this blur of experience, a concept that takes account of everything happening in the mind yet remains firmly entrenched in the present moment. All that is happening in the mind at one instant, all the firing nets of sensation and memory, can be called our plane of consciousness—or our conscious plane.

The conscious plane is a global term to describe everything we are aware of, everything we feel, and everything we are thinking about at a particular moment. Because it is built out of nets, it should share most of a net's properties, so, for a start, the conscious plane will be self-defining. It is not a skull-sized bucket waiting to be filled with all life's experiences but is only as large as the areas of brain that happen to be active at a given point in time. When we are brightly awake and feel powerfully conscious, it is because our brains are active with lots of nets and the conscious plane is painted large. But when we are drowsy or asleep, what we define as the conscious plane has shrunk to almost nothing because few nets are firing. The conscious plane stretches elastically, depending on how many nets a hardworking brain can juggle at once.

The nature of nets is behind another property of the conscious plane—the way the conscious plane feels as if it is made up of an endless smooth flow of mental events. We have seen how individual nets have no sharp boundaries. While a net may have a definite center, it is blurred at the edges where it merges into the gray background of surrounding nerve tissue. But nets can also be said to blur into time because of the way they first burst into life like a flaring match and then burn back down again to fade gently from consciousness. Because the conscious plane is made up of many nets, it is formed by a mixture of nets that are at different stages of this process of birth and decay, which gives our consciousness an illusory feeling of flowing permanence. Nets do not arrive and depart abruptly but rather seem to swim into view and then slip away almost unnoticed as other nets slide in to take their place. So the coming and going of nets is smooth enough to give us the false feeling that our conscious plane sails serenely on, independent of the many fleeting mental events that make up each separate instant of our awareness.

Having defined the idea of the conscious plane as the sum

of all the nets active at a particular moment—and having also described perception, recognition, and memory in terms of nets—perhaps we should pause to check our bearings. We have so far looked at the working of the brain only through human eyes, and have ignored animals. It is time to bring the evolutionary perspective back into the picture and see if the animal mind works in much the same way.

THREE

Rousing Memories

Most biologists shy away from speculating what it must be like to see life through the eyes of an animal. They stick strictly to observing an animal's behavior and do not concern themselves with its conscious experience of the world. But we need some understanding of the natural mind of an animal if we are to see how the human mind differs.

The great physical similarity of the brains of higher animals, such as birds and mammals, and man should show that our minds work in a broadly similar way. Perception would certainly seem to be much the same for a cat or kangaroo as it is for us. Animals have sense organs like ours, a cortex mapping of the world like ours, and a rising froth of impressions like ours, from which we would expect animals to be conscious of the world like ourselves. The mapping and processing of the world on the cortex is what creates raw consciousness in humans, so, given that the sensory and neural

equipment of animals is the same, we can expect all verte-
brates to show at least some degree of consciousness.

An obvious difference between the human brain and most
animal brains is size. Because the human brain has a tablecloth spread of cortex for sensations to dance across, compared with the handkerchief of an ape or postage stamp of a
rat, we would expect human consciousness to be richer and
sharper simply due to the extra neurons that can be devoted
to mapping the detail of the world. But as we have said, the
forming of raw impressions in the mind is only half the story
of consciousness: To feel truly aware, we need to understand
what passes through the mind. We need to forge a connection between sensation and memory; otherwise we would
experience only a chaotic kaleidoscope of impressions. Animals can certainly do this too. There is no doubt that they
have good memories for recognizing the familiar: All higher
animals can recognize the cries of their children or the faces
of their mates; they can find their way back to their burrows
or learn a complex path through a maze.

The memories of animals are often at their most impressive when it comes to food. Marsh tits hide caches of food
through the day and can store several hundred grubs and insects to be collected up to a day later. Chimps also show very
good memories. In one experiment, a scientist carried a chimp
around while he hid eighteen pieces of fruit in an acre-sized
compound. When he let the chimp free an hour later, it could
remember its way back to about a dozen pieces—running
straight to its favorite types of fruit first. Then, after eating
most of the fruit and resting a few hours, it suddenly got
up—presumably prompted by fresh hunger pangs—and ran
off to gather some more, showing that the memory must have
persisted.

A purer test of the powers of animal memories can be made
by teaching an animal to associate the showing of a colored

card with the arrival of food a short time later. As might be expected, the duration of such memories was linked with brain size. If there was more than ten seconds' delay between card and food, goldfish would forget the significance of the signal, whereas pigeons and lizards could remember for two or three minutes and baboons would wait for the food to arrive up to half an hour later. However, the important point is that all animals—even the humble goldfish—show at least some memory ability.

Since animals have perception, recognition, and memory, they should also have a conscious plane like ours; they should also have "minds." Their minds will have less detail, less sharpness, and smaller memory capacities because most animals have much smaller brains on which to map the world, but these are differences in quantity rather than quality. We should expect animals, like us, to be consciously aware of their surroundings, to open their eyes and have impressions of the world come flooding in, striking sparks of understanding in their memory banks.

It is obvious that the human mind is different in some important way from the mind of even a close relative like the chimpanzee. However, the difference is not in the neural foundations that have painstakingly been laid down during millions of years of mental evolution but in what humans have recently been building on these natural foundations. We shall see how language has been the tool with which humans have sculpted extraordinary new mental structures on top of the mind's natural foundations. Some of the new structures are obvious, such as self-awareness and rational thought. Some are far less obvious, like complex emotion and personal memories. But all these structures are the result of humans finding new ways to use the existing hardware of the basic animal brain and not the result of any fundamental break-through in the way that brains are made. Neurologists have dissected the brains of frogs and sharks and found that, so far as they can tell, the brain cells are no different from a human's.

Animals may be consciously aware, yet it seems a blank kind of awareness. If we were able to live briefly inside their heads, we would find the animal mind strangely uncluttered. There would not be the same churning of past thoughts and future plans that fill the human mind. There would not be the continuous chatter of our inner voice, nor the sudden breaks to reconsider our own actions as we switch from simple awareness to self-awareness.

A cat is an example of the clear uncluttered animal mind. Anyone who has had one as a pet will know how sharp and alive a cat can be, yet also how often oddly blank. For the feline mind, life seems to be a series of moments strung together, with no inner turmoil of thoughts or concerns to ripple its placid surface. From what we know of the way a cat's brain processes sensory information, we can assume that when a cat spots a mouse scurrying across the lawn, impressions of the mouse will fill its conscious plane. The cat will recognize what it sees and this recognition will awaken memories of what to do. The cat will excitedly start stalking and trying to pounce on the mouse. However, if the mouse is hidden for a short time, the memory of the sight will fade after a few minutes and the cat will wander off. Out of sight will be truly out of mind for the cat. If the cat later passes the spot where it saw the mouse and catches a whiff of its scent or recognizes the corner of the lawn as having previously had a mouse on it, the hunt for the mouse may be on again. But it is important to note how the environment is in control of what is happening in the cat's mind.

As should become clear later in the book, lack of language is the key to this empty-sounding moment-to-moment existence. When nothing exciting is happening, a cat will sit blankly. It will not be cursing itself for missing the mouse or lazily daydreaming about a fat mouse it caught a couple of weeks earlier. The cat is aware, but its awareness will be filled with impressions of the moment—like the warmth of the sun, the twittering of birds, and the scents of the garden. It has no mechanisms for riffling through its memory banks at times of leisure or even for considering the fact of its own existence. Life is lived in the present tense. Only memories connected with the events of the moment become aroused enough to form part of the conscious plane.

By the same line of reasoning, we can argue that even highly social mammals, like dogs and apes, live only in the present. Chimps may often look as if they are thinking even when simply sitting in the shade of a tree, yet they are still being driven by the changing world around them rather than responding to chains of internal thoughts. The only difference is that chimps have far more going on around them than do solitary animals like cats. They are part of a group with complex social relations and so need constantly to keep an eye on what other chimps are getting up to. This social pressure will trigger a steady train of nets in their minds as they recognize familiar situations while watching the activity of others with the seasoned eye of experience.

It is very hard to imagine being inside the head of an animal, but the essence of the difference is that animals are chained to the present. Their wordless minds can react only to the events that surround them at a particular moment. Human minds, however, have broken free. We can think about the past, make plans for the future, and fantasize about imaginary events. The key advance humans have made is that we have a measure of inner control over the thoughts and visions that flicker through our awareness, while animal minds are shackled to the world around them. To find out where this inner control comes from, we must take a closer look at how our memories and imaginations work.

We can start by drawing the distinction that psychologists

commonly make between working memory and long-term memory. Working memory—or short-term memory as it is often called—is the amount of information we can keep in the forefront of our minds at any particular moment. It is the memory capacity we have to play with when trying to remember a list of words, a drinks order, or a long phone number.

Information kept in this working memory is easy to grab hold of as it appears to hang around on the fringes of awareness. Everything that passes through the mind spends some time as a fading echo in this temporary parking space. When we glance at one corner of the room and then turn away, a fairly fresh impression of what we have just seen will linger in working memory. Or if we taste a piece of chocolate cake, smell fresh orange juice, or feel an ice cube, the sensation will hang around for a short while. Even thoughts and emotions last for a time in this working space.

Exactly how long impressions and thoughts linger is hard to measure, but it seems that most things last for at least ten seconds or so—unless there are many distractions to shorten the time or we have gone back to refresh the impression. This sort of recall is for the type of pinpoint-sharp memory of recent sensations sometimes called iconic memory, where the memory feels almost as fresh as the original sensation. A step back from this comes another level of rather less intense working memory which lasts for about ten minutes. Impressions seem to linger here, close at hand, but not quite so fresh. The recall is more forced and blurred. This hazier working memory eventually fades away to leave only long-lived traces etched in long-term memory.

Before looking at how impressions come to be "fixed" in long-term memory, it is important to note that the limited life of working memory is a very real restriction. Some people who have suffered severe brain damage because of clots caused by strokes find it impossible to store lasting memories; they are fully aware of the present but forget everything they have just experienced after about ten minutes. One such man kept saying he felt that he had just waked up or that he had been dead until a few minutes before. He could remember enough in working memory to keep going from moment to moment but a terrifying blankness chasing up behind him left him feeling as if his whole life was restarting every few moments.

It is hard to imagine living with a blank wall cutting ourselves off from our pasts, but some of the limitations of memory are clear from the way we struggle to remember our dreams. When we are asleep, the mechanism that fixes long-term memories is obviously switched off—or else our dreams do not even feed into it. To remember dreams we have to wake up and replay the sensations that still linger in working memory so that they will be properly fixed. If we do not do this immediately on waking, the dreams are lost to us forever.

The capacity of working memory is even harder to gauge precisely than its duration. It is certainly limited. In psychological tests, subjects are read a list of nonsense words, letters, or numbers to find out how far their short-term memory will stretch. Most people can remember only about seven chunks of information—seven words, numbers, letters, or whatever—and even exceptional people can manage only nine or ten. This limited capacity is obvious from everyday experience. While we can look up a local phone number in the directory and remember it long enough to dial, we usually need two attempts at remembering the long code of an international call.

These sorts of tests of capacity give the impression that we have a mental workbench on which we lay out all the memory chunks we want to work with. Any more than about

seven chunks will not fit on the table top, so they are either
sent away to be stored in the cupboards and shelves of our
long-term memory or else they are shoved off the edge of the
table and forgotten. But the idea that working memory has a
fixed capacity is misleading. The capacity is elastic and can
vary with how alert we are feeling. Also, while we may
struggle to keep seven chunks at the forefront of the mind,
we find it easy to hold two or three. Just like other important
mental abilities, such as vision and nets, working memory
has no fixed boundaries, only a blurred fringe. The reason for
this is that working memory is really another way of looking
at the conscious plane.

As we have seen, the conscious plane is made up of all the
sensory and memory nets that are firing at a particular moment. Working memory is another way of describing how
nets linger awhile in consciousness before slowly burning low
and fading into grayness. A net bursts brightly into life when
it comes into the center of conscious attention; then, as the
spotlight of attention shifts to a new net, the old net joins
all the other spent nets huddled on the fringes of awareness.
Eventually the edge of the conscious plane gets so crowded
with used nets that they tumble over and disappear. The capacity of working memory is defined by the number of nets
we can keep alive and glowing in consciousness at any one
time. If we have just a few nets on the go, there is plenty of
mental elbow room and they can burn on in the background
of awareness for quite some time. But if many nets are created, the conscious plane becomes like an overloaded life-
boat and some nets will be tipped overboard.

Working memory may be just another way of looking at
the conscious plane but it is a useful concept because it lends
an extra dimension to our picture—time. We have already
talked about how the workings of the senses and memory
banks create the dancing patterns of consciousness, but this

describes only how nets are physically arranged in space. Working memory brings in the effects of time. The sharp focus of consciousness is not just the center of a net but also the center of the net that happens to be the most brightly burning net of the moment. The blurred fringes of consciousness are not only the outer boundaries of nets but whole nets that are either just firing up or gently winding down. When we take the two dimensions together, we can see how the conscious plane can lack clear boundaries in either space or time and yet still have a sharp central focus of awareness.

The added element of time also allows us to build up a picture of the world spread over some seconds rather than having to rely on snapshots of the present moment. If most impressions of the world tend to linger for about ten seconds, we can see life as ten-second newsreels rather than a series of still frames. The eye can flick across a scene, picking up more detail through new points of focus. Working memory then keeps this series of snapshots jangling long enough in awareness for a full mental picture to be built up. The mind can get to work on recognizing this picture, mobilizing the nets of memory needed to understand and act on what is seen. Working memory describes how the mind, like a juggler, keeps all the balls of memory and sensation in the air long enough to do some useful thinking.

An important point about working memory is that this blurring in time not only extends into the recent past but also reaches forward into the future by half rousing memories and half focusing on fresh points of interest, thus dragging the mind along to new things.

Most people have only a very hazy feel for what drives their minds on to the next thought, image, or memory. Their minds seem to charge along of their own accord with no visible means of propulsion or steering. We assume that we are in control until we ask ourselves why we came up with this impulse or how we worked out that idea; then we can begin to feel like passengers being taken along for the ride. This puzzling momentum gives weight to the feeling that the mind is driven along by some subterranean force, like the Freudian unconscious.

However, the mind is like a stage. Nets have to step out of the wings and cross to center stage to come under the spotlight of attention. This can be seen from the way our eyes are often dragged to their next focus of attention by a sharp movement out in the blurred fringes of our field of view. What exists only in the corner of our vision is, in a sense, our future. The next sudden movement out in the broad margins of sight is guaranteed to be our new focus of attention. The same applies to the other senses. The noisy starting up of a neighbor's lawn mower or an itch on the leg begins on the blurred fringes of awareness and then rapidly thrusts itself into the center of our attention.

We usually overlook the way these fresh nets start off out at the fringes. This is hardly surprising because—by definition—we are always sharply focused on the images and memories at the center of attention rather than worrying about what might be coming next. Indeed, things often happen so swiftly that they have passed by the time the spotlight of awareness arrives on the scene. For example, we might be surprised by a ringing gunshot during a quiet afternoon at home. The noise is over by the time we turn our full attention on it, so we have to focus on the echo, which fortunately still lingers in working memory. And perhaps when we examine the noise properly, with the benefit of our sharp attention, we might find that it was only a car backfiring. Attention is constantly dragged along like this by life. Our eyes and ears are caught by something half sensed but interesting, which is then brought into central focus so that it can be properly recognized and understood. By this simple mechanism of chasing the fringes of awareness, our minds are whisked along at breakneck speed.

The half-seen fringes of awareness can drag our thoughts along just as they do our attention. As with the fisherman above the net of memory, each handful of net lifted into bright sunshine also pulls a knot of closely linked facts nearer the surface. We cannot quite see what the facts are when they are still just submerged, but we can see enough of them to be tempted to reach in and lift a new handful of memory to the surface. This is important for the association of ideas that drives along our chains of rational thought. We hop from idea to idea as if we were skipping across stepping-stones in the mind; with each new step, we half arouse several more potential footholds and leap toward the half-seen idea that looks the most interesting.

The associative power of the memory surface is important in simple, everyday types of thought. For example, seeing a rabbit hole while out walking should bring forward the mental image of a rabbit. The two memories are close together on the underlying net of knowledge, so experiencing the sight of one will gently tug the other into conscious glowing life. This is useful because once we have caught sight of a rabbit hole, our eyes will be on a hair trigger to spot a rabbit—even one crouching almost invisible in the long grass. With our rabbit knowledge dragged halfway into awareness, it is easier to recognize the real thing.

If we were hunters out looking for wild game, this natural association of ideas would also awaken a whole set of knowledge about how to catch rabbits and the likely startled response of the rabbit if it saw us. By a rousing of memories, we would be forearmed with all the knowledge needed to make the best effort at catching our dinner. This example of thoughts taking place in working memory is one that probably applies as much to the mind of a fox as to a man's. Perhaps a fox might not have the intelligence to recognize a rabbit hole or spot muddy rabbit footprints on the trail, but a fox could smell the fresh scent, which would spark a similar awakening of stored rabbit memories to make the fox a better hunter.

Working memory is often talked about as if it were separate from awareness, a convenient scratch pad used by our consciousness. But as we have seen, it is part of the process of being conscious rather than some backup mechanism. It is not a special patch of brain but rather a description of the mental elbow room provided by the way nets linger before fading into grayness. Having seen how nets get crowded to the edge of working memory and disappear, we can now look at how much eventually gets stored in what we call long-term memory—and how much of life's experiences are lost forever.

Long-term memory is even harder to measure than working memory. Every person's head seems to contain enough facts, even if trivia, to fill several encyclopedias at least. We also seem to store an enormous wealth of personal memories spanning the years right back to our childhoods.

Some idea of the limits of long-term memory and how it works is given by studies of people with extraordinary memories. One of the earliest reported of such cases was a nineteenth-century Russian reporter named Shereshevskii who astonished his editor by never taking notes during briefings but being able to repeat the detailed instructions word for word. His editor sent him to the Russian psychologist Aleksandr Luria, who found that Shereshevskii could remember, with little trouble, lists of hundreds of numbers, long strings of nonsense syllables, and poetry in foreign languages. He could even repeat the test pieces backward and remember them accurately after a gap of several years.

The secret of this extraordinary memory was the intensity

of Shereshevskii's imagination and the way sensations tended to spill over and affect each other inside his head. For example, the sharp chime of a bell was not only experienced as a ringing noise but also sparked flashes of light, tastes, and feelings in his mind. He described one particular tone rung by Luria as looking like pinkish fireworks with a rough texture and an ugly taste of briny pickles. The voice of a friend was described as yellow and crumbly.

This sort of spillover—known as synesthesia—is present in a mild form in most people. We might, for example, describe music as sweet, as if we could nearly taste it. But Shereshevskii had an extreme and uncontrollable reaction to everything he came across, making unremarkable details of the world, such as a list of numbers, stand out memorably in his mind. To Shereshevskii, numbers had shapes and colors: The number two, for example, was flat, rectangular, and whitish, or was a high-spirited woman. When he was faced with memorizing long lists of numbers and words, each digit or letter would provoke powerful images that Shereshevskii could work into a little story. One complex mathematical formula, for instance, involved a shriveled-up tree, houses being built, and a harmonica-playing man standing by a post office, with each fragmentary image representing a mathematical notation. Even fifteen years later, Shereshevskii could remember the strange little story as if it had been a major event in his life and so recall the formula quite easily.

Even without such stories, the impression left by a relatively featureless string of numbers was so strong that Shereshevskii could repeat them as if reading from a list he had chalked up on a mental blackboard. Many people facing exams might wish for this sort of memory, but while Shereshevskii eventually made a living as a professional mnemonist, his extreme reaction to sensations was as much a handicap as a blessing. Luria described him as a timid and ponderous person who was weighed down by detail and often had trouble understanding what he was remembering. Shereshevskii complained that every word he heard sent a chaos of images tumbling through his mind, so that he could not follow the sense of complex sentences. Common metaphors such as "weighing one's words" simply left him confused. His detailed memory created other problems: He had difficulty in recognizing people because their faces never looked exactly the same as the last time he saw them; and if someone coughed while he was trying to memorize a list, the cough would be smeared across his memory when he came to recall the material. Indeed, Shereshevskii's imagination was so powerful that he had often been late for school because he would imagine that he had already got up and been to school that day.

Shereshevskii, although an extreme example, is only one of many people who have shown this sort of extraordinary memory. But his ability to recall long mathematical formulas by using imaginative stories is not the same as having a photographic memory for events. A photographic—or iconic—memory is the ability to hold crystal-clear images in the mind for some seconds or minutes, rather than years. Because these mental pictures are so vivid, it can feel almost like looking at the real thing.

Although fewer than one in a hundred adults in the West have this sort of intense imagery, it is quite common in children and "primitive" people. In a 1960s study in a Nigerian village, a slide projector was set up and the tribesmen were shown some pictures for thirty seconds each, ranging from a photograph of a Nigerian bus stop to scenes from *Alice in Wonderland*. Over half the villagers showed some level of photographic memory and about a fifth had almost perfect recall, being able to do such things as trace out the license-plate number of a car from their memory of a picture, even though they were unable to read or write. In one instance, a subject who wrongly stated that the Cheshire cat from the *Alice in Wonderland* picture was black was greeted with cries of scorn from the other eighteen villagers who had watched the test. All of them were looking at the blank projector screen as if the picture still lingered there like an after-image from staring at the sun, and when they were asked how many could see the original image, fourteen hands shot up.

These impressive memory feats—even though the mental pictures quickly fade—seem, however, more the sign of an uncluttered mind than of special powers of memory. Members of the same tribe, brought up in cities and educated to read and write, show far less ability, and studies in the West have shown that while eight in a hundred children appear to have photographic memories, nearly all lose their ability as they become adults.

The reason for this loss of natural photographic ability has not been explained, but it could well be the price paid for being taught from an early age to analyze and label all we see. We establish filters that break up sights and sounds as they enter working memory, which probably makes it difficult to hold complete scenes intact in our minds in the same way that the naïve mind of a child can. Our educated eye fragments the world around us. However, by filtering out irrelevant details and searching out patterns, we are perhaps left with a more thoughtful reaction to what we observe.

This brief look at some of the extremes of memory should show that how much we can potentially remember depends largely on the type of processing that goes on when we first take in the information. But is there anything to the common feeling that all life's experiences are automatically stored away somewhere in the brain—like junk in a huge attic—and we need only the right key to get them back? This seems true from the way an old diary can bring long-neglected memories flooding back as vividly as if the events had happened the day before. Also, our apparently unlimited ability to recognize pictures and faces seems evidence that all the original experiences left some permanent imprint on our memory banks. To answer this question properly, we need to go back to the basics of memory and look at the ability from the evolutionary perspective.

Simple animals, like slugs, worms, and insects, lack memory as we know it and instead depend greatly on instincts to control their behavior. The way a fly is startled by movement, or a spider knows how to build a web, is built into the wiring of their nervous systems by millions of years of evolution. They do not have to experiment and learn. These instincts have become "hard-wired" or genetically programmed over many generations.

However, even the most primitive creatures show a rudimentary level of learning, so these instincts can be fine-tuned. This is shown by experiments that have been done with an ocean-living relative of the slug family called *Aplysia*, which has only about ten thousand nerve cells in its six-inch body, compared to the billions in more advanced animals. The most basic form of learning in *Aplysia* is the deadening of the simple reflex reactions wired into its body. A prod from an experimenter's finger normally makes the sea slug hastily pull in a frond of feathery gills on its back, a reflex that prevents the gills from being bitten off by passing fish. But when the poking is kept up for a few minutes, it is eventually ignored. The nerves carrying the alarm-bell messages from the skin to the small muscles controlling the frond become deadened or "tired" by the continual stimulation. The deadening is a useful reaction since it allows reflex circuits to become tuned to an animal's surroundings, so if, for instance, a sea slug found itself in a rough sea one day, it would not spend the whole time curled up in a ball.

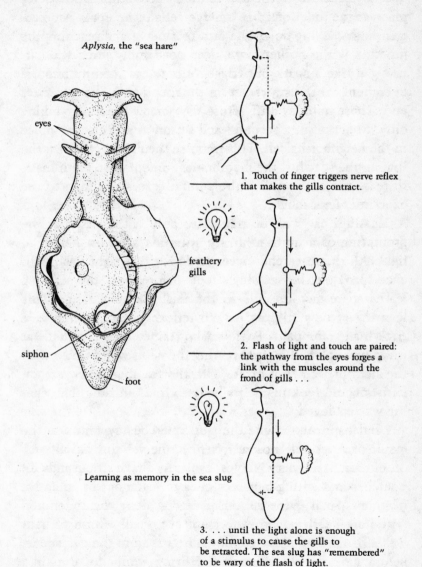

Aplysia, the "sea hare"

eyes

feathery gills

siphon

foot

Learning as memory in the sea slug

1. Touch of finger triggers nerve reflex that makes the gills contract.

2. Flash of light and touch are paired: the pathway from the eyes forges a link with the muscles around the frond of gills . . .

3. . . . until the light alone is enough of a stimulus to cause the gills to be retracted. The sea slug has "remembered" to be wary of the flash of light.

This deadening of the nerves to repeated stimulation takes place at the junction of two nerve cells or neurons. An electric pulse passing down the arm of a nerve and reaching the junction with another nerve does not simply jump the narrow gap like a spark but triggers the release of tiny bubbles of a chemical transmitter from the membrane at the nerve's end. These bubbles drift across the narrow junction—taking only a fraction of a second—and set off a new electric pulse in the neighboring cell. This impulse then races off down the line. In the sea slug, the membranes run out of steam if made to react too often and so the animal ceases to flinch to the experimenter's touch.

Sea slugs can also be trained to pull in their gills at the prompting of a quite arbitrary stimulus such as a flash of light. All the researcher needs to do is pair a light flash with every prod on the sea slug's foot. Before long, the sea slug learns to retract its gills to the flash of light alone. This learning of new triggers for old reflexes is a commonplace trick made famous by Pavlov, who trained dogs to dribble at the sound of a bell by first pairing the bell several times with the sight of a bowl of meat. With the sea slug, however, scientists could look inside its body to find out how this leaning was achieved.

What happened inside the slug's nervous system was the result of a natural tendency of firing nerves to link up with other firing nerves. Nerves typically have thousands of branches and will sprout new ones to reach out to an active neighbor. So if every time the nerves along the touch and withdrawal reflex loop were being triggered, a flash of light was also sending messages down nerves from the eye stalks, before long the two patterns of firing would forge a link. Connections would sprout where none had existed before, and with continued use, these new connections would thicken and strengthen. The longer the pairing of touch and light went on, the stronger the bridge between the two active nerve pathways would become until, eventually, the light flash alone would feed enough nervous stimulation into the reflex pathway to trigger off the retraction of the gills. The sea slug would have "learned" to flinch to the unlikely stimulus of a flash of light.

It could be argued that this simple rewiring is the basis for all learning and memory in higher animals; that the tendency for active nerves to forge links is the key to everything. The sea slug with its very simple nervous system has only limited scope for such creative rewiring but as animals developed, they grew large brains with billions of nerve cells clumped together and evolved the ability to represent the outside world as a pattern of nerve firing splashed over the surface of the brain. Once animals had reached this level of sophistication, the fleeting sensations of life could leave a permanent trace.

As we saw before, when considering the way the brain sees a cow standing in a field, a network of neurons in the cortex is set jangling by stimulation from the eyeballs. Our visual sensation of seeing a cow subsides soon after we turn our eyes away; the retinas stop sending cowlike messages to the brain and the jangling network that was whipped fleetingly to life dies gently away. But if this burst of firing is followed by chemical changes of the same kind that resulted in simple learning in sea slugs, then the connections briefly forged by the aroused net could become fixed. The individual cells might fade back into grayness or get involved in new nets, but a pattern of connections would have been etched between the cells, leaving a faint imprint. When this clump of cells is later fanned into flame again, the reawakened net would cause something like the original cow sensation to burst back into conscious awareness once more.

It would not even be necessary to stimulate the whole net

of neurons again to get such a response. Because of the links
tying the network of cells together, a small tug at one corner
of the net could trigger a chain reaction. If we had been shown
a photograph of the cow, perhaps just a square inch out of
the center of the photo would be enough to bring the complete image back to mind. Or because humans have language, just saying "cow in a field" would set the stored net
firing again.

The exact nature of the chemical changes that form the
bonds of long-term memory is still uncertain. They would
have to be fast-acting and long-lasting changes but also leave
neurons relatively free to take part in other nets. Neurons
that might one second be representing part of a black patch
on the cow's back might the next second, with a shift of
gaze, have to become the dark hollow of a tree stump in another corner of the field. Some neurologists have singled out
the enzyme calpain as being important because, when it is
blocked from doing its work at nerve junctions, rats stop
learning in laboratory experiments. But hundreds of brain
chemicals have been discovered, and calpain is likely to be
just one link in a chain of chemical reactions that govern the
firing patterns and sensitivity of neurons.

This variety of control over nerve activity gives a basis for
long-term memory. However, while everything that enters
conscious awareness should apparently leave some imprint
on the mind, this does not mean that the mind keeps a photographic record. The daily torrent of life's experiences is not
simply channeled into a vast reservoir in the back of our heads
in which we then have only to fish to recover what we originally saw. Memory is an active process and experiences are
filtered and twisted while passing across the center stage of
working memory. When the firing nets of nerves collapse into
the grayness of the surrounding brain tissue, their fate depends to a large extent on the shapes and patterns already
etched into the memory surfaces.

For instance, a powerful and unique experience—like being involved in a car crash—will leave its own stamp on the surface. While most of the neurons involved in originally viewing the crash scene might frequently play a part in other sensations and memories, the intensity of the crash experience would burn in the net. The neurons would strengthen the bonds between them so much that even a gentle tug at one corner—such as the daily act of getting into a car—might set the whole nasty memory firing through consciousness again. Such a memory will also be helped by its uniqueness. Unless the driver is a stock-car racer, he is unlikely to have experienced other crashes, so the crash memory will not become overlaid or encrusted with the details of later smashes.

The same cannot be said for the commonplace events of life. Perhaps you can vividly recall drinking a cup of coffee yesterday, but coffee-drinking memories from more than a few days ago would all congeal into a blurred general net of ideas. While you might remember that you generally use a favorite mug or that you switched brands some years ago, and you could glean enough details from this blurred net to reconstruct a typical coffee-drinking episode, it would be hard to swear whether any particular detail was 100 percent accurate.

The trace that is left by an experience, then, depends on whether it blurs into a general body of knowledge or has the uniqueness and power to etch its own pattern. Despite what we commonly believe about our memories, the trace that is left is not a complete and detailed record. The brain was shaped by the economics of evolution, so memory was designed to do its job with as few energy-wasting frills as possible—and that job was to recognize the familiar and to create a general backdrop of knowledge about the world.

The evolutionary advantages of being able to recognize the familiar are obvious. Long-term memory allows the higher animals to identify and associate ideas with the important

things around them—such as fellow troop members, their own
burrow, or the sight of predators. While we humans emphasize our visual memories, most higher animals are at least
equally dependent on the recognition abilities of their ears
and sense of smell.

As well as permitting recognition of specific things in the
world around us, long-term memory has the second, equally
important job of giving animals an internal "mental backdrop," so that all new experiences can fit into a context of
understanding. For instance, as it grows up, a mouse builds
up a general picture of the world so that every day that passes
its pool of knowledge increases about the way corn smells,
predators look, or branches bend. This inner picture of the
world means that when a problem crops up—such as getting
at an ear of corn—the mouse already has background knowledge about swaying stems and swooping owls.

We tend to overlook the importance that this general world
picture has for human thought. It is so fundamental to our
thinking that its existence is easy to miss. For instance, the
knowledge that apples will fall to the ground when dropped
or that thin branches are likely to bend under our weight is
taken so much for granted that we never wonder where it
comes from. It can be very revealing when some of our com-
monsense ideas about the world turn out, in fact, to be quite
wrong. For example, Galileo caused an uproar in the six-
teenth century when he pointed out that two balls of different weight would both hit the ground at the same time. Our
natural guess would be that the heavier ball would fall faster,
and even when we are given an impressive demonstration—
such as a feather and a lead ball falling at the same rate in a
vacuum chamber, with no air resistance to slow the feather—
it is hard to rid ourselves of the intuitive picture of what
should happen. Another similar commonplace misconcep-
tion is that a ball falls straight to the ground if dropped while

a person is walking along, whereas, in fact, it falls in a gentle curve because of the forward motion of the walker.

Such fundamental errors about the way the world works should shake any lingering belief that the mind is an infallible mirror of reality. However, these sorts of mistakes about the motion of objects did not cause any real problems for humans until the Middle Ages when people started to fire cannons and build machines. They would certainly not have bothered early man, who, with the size of his brain, was already enjoying enough of an advantage over other animals in understanding how the world worked. Of course, the greater the accuracy of perception, the better, but evolution never needed absolute perfection to get brains to do a useful job.

Continuing with this evolutionary viewpoint, the first *Homo* species clearly did not need the sort of personal memories that we feel are so important to us humans. They would have been well enough served for evolution's purposes by their general ability to recognize the familiar and to mobilize a backdrop of understanding. Like a fox hunting rabbits or a cat stalking mice, our ancestors would have lived in the present tense and used memory only to broaden their understanding of what was going on from moment to moment. Modern man's memory for his own personal history is an artificial addition to the mind, made possible only after he had developed speech. To see this, we should consider how we recall what has happened in our lives.

Recognition, as has been described, involves the powerful prodding of a repeated sensation, fanning fresh life into old pathways. With the tingle of the "aha" feeling, we can recognize thousands of photographic slides that we may have seen only once—and for just a few seconds. However, when asked to recall as many of the pictures as possible, most of us would be able to describe only about a dozen of the thousands of slides. Conscious recall is not only far weaker than recognition, it also feels different. Recognition just happens to us; we see something and we recognize it. The work is all done behind the scenes by the natural workings of the brain. But recall involves an active scouring of the memory banks.

If asked to recall some of the thousands of slides from the test, we would resort to a host of little tricks to jog our memories. Half a dozen slides might spring to mind quite easily— as many as might linger in the limited space of our working memories—but after that we would start using such tricks as imagining a blank screen and running through a list of likely items that might have come up on it. For instance, we might mentally suggest flowers, and after we had pictured, say, a few roses, it might hit us that a dandelion was one of the images that we had seen. The picture would suddenly come vividly back before our mind's eye, as if thinking about roses brought us close enough to the net where dandelion memories were stored for the spark of memory to leap the gap.

Conscious recall involves all sorts of such ploys to root out buried memories. We might go through the alphabet to see if each letter triggered any images, or we might try to think what sort of pictures were likely to be included in a psychology experiment. So, apart from the photographs that come back immediately from working memory, we are faced with a difficult conscious problem, not a swift clean response like recognition.

Another example of how we dig out our personal memories is the way we would try to recall all the names and faces in our school class at the age of, say, fourteen. The mental strategies we would use, as with the slide sequence, would probably be a mixture of logical searches and attempts to imagine the context as vividly as possible. We might picture a sunny summer's afternoon with Mr. H., the math teacher, scratching chalk on the board while we chewed a pencil at the back of the class. As we looked around this imagined classroom, clusters of faces should come back to us—our group of best friends, the class grind, the class big shot. Or we might try logical tricks such as mentioning a few names to see if they jog memories, asking ourselves if there was a Sarah or a Tom in the class, or else working out that fourteen is when we started playing hockey so there must have been some of our teammates in the room.

The way we went about remembering our school days should show how we try to re-create the ground swell of sensations that would normally lead to recognition. We try to imagine the original scene or else we use words to take us near enough to lost memories for a spark to jump the gap and light up a long-forgotten face. Either way, recognition is at the bottom of our conscious recall. We construct elaborate chains of thought which we hope will lead us very close to where the memories are lying on our vast fisherman's net of memory. Having got close enough, we can let our natural recognition processes take over.

This detective work relies on an inner control that comes only with language. When searching out buried memories of school friends or dandelion pictures, we use an inner dialogue to create an imaginary context that will bring such memories naturally to mind. We then use language to fashion these awakened traces into a re-creation of events. We prod our memory banks with questions and fire our imagination with words.

The power of speech in our minds is obvious from the way a single word can strike a wealth of associations. For instance, the word *picnic* will cause our mind to echo with picnic memories: We may have a storybook image of children sitting around a checkered cloth or remember a picnic hamper full of plastic cups. A single word can lead us to a buried network of related ideas that then pour forth into consciousness as if we had turned on a tap.

There is nothing magical about the way words have this

power. Words are only noises and the brain processes them just as it does all other noises, so that the sound of a word is mapped onto the hearing zone of our cortex and becomes a conscious sensation just like the whistling of the wind or the ringing of a telephone. The difference from these everyday sounds is that the brain has been trained from childhood to link the sound of certain words with certain patches of memory. Nerve pathways are forged as we grow up so that when the noise "picnic" is registered in the part of the brain that hears, jangling messages are triggered which cross the brain to wake up the patches of cortex where picnic-type memories are stored.

In this way, words give us a second route to our warehouse of memories. In animals, only the bottom-upward stimulation of sensation can trigger stored memory traces. They need the reminder of familiar objects in the outside world to bring back memories about those objects, so they are always shackled to the present. But with speech, humans have a second, top-down, route to memory. Because speech is a skill under our own control, we can steer our consciousness around our vast memory surface, exploring the past or wondering about the future.

The effect of words on the process of perception itself should also be noted. We perceive the world much as animals do, the brain being built to filter and sharpen the images captured by our senses and to deliver them to the clutches of memory for interpretation. But our extra ability to name what we see makes a big difference to human perception because of the control it gives us over this natural process.

Naming something makes it stand out more clearly from the surrounding background. For instance, if we wordlessly scan a familiar room, we see everything in it, yet nothing has any particular impact. An animal would always react blankly to the familiarity of such a room and take special notice only if something dramatic was happening—a whirring toy in the corner perhaps. Humans, however, can focus on one object and name it—a chair or a curtain, for example—and it is as if we have clipped it out of a picture with a pair of mental scissors. When lifted clear of its background, the chair or curtain is no longer a vague feature of the landscape but a particular memory, neatly trimmed around the edges and separated from its original context.

This word-sharpened perception is what allows humans to focus on and think about anything we sense—even something as slippery as our own thought processes. Animals have their thoughts thrust upon them by the filters of perception. Their brains have evolved to echo the most important happenings in the world around them, so their awareness will always lock on to the attention-grabbing features of their surroundings. What an animal focuses on is dictated by the environment. It has no mechanism to decide for itself what it wants to concentrate its awareness on. But humans, by using words to clip out bits of the world with mental scissors, can deliberately focus attention on parts of the world that, superficially at least, appear boring to our brain pathways.

Words are behind another of man's artificial abilities—imagination. Our fantasies or imaginings rely on the same language-driven mechanisms that allow us to relive memories, the only difference being that we know our imaginings are deliberately fictitious.

Once we have tagged a fragment of the world with a word, we are also able to shuffle perceptions around in our head. Using the words as place markers, we can assemble a mass of detail to create imaginary scenes. We use the words as handles to pick up sets of ideas and put them down together on the brain's visual surfaces to see what sort of picture results. For example, the phrase "Imagine a blue crocodile" should prod our brains into a sharply focused response, with the words "blue" and "crocodile" leading us to two well-defined sets of ideas and tugging our stored knowledge out into the open. We could then set to work to visualize a blue crocodile by merging the two sets of ideas in the theater of our working memory. We might start by picturing a typical crocodile lazing on a muddy riverbank—but its scales will be muddy green and we will have to try mentally shading them blue. This might prove difficult, for our net of knowledge about crocodiles will be fairly certain about their proper color and even after considerable effort we may well fail to put a convincingly blue picture before our mind's eye. The strength of our imagination is limited by how far we can stretch the bundles of ideas tagged by the two words, *blue* and *crocodile*. However, we may find extra words to give a more flexible framework for our imaginings. Perhaps we think about a cartoon crocodile—maybe half remembered from Walt Disney's *Fantasia*—and suddenly we have assembled the right set of ideas to allow us easily to imagine a crocodile in blue, or indeed in pink or yellow, dancing in a tutu or singing a song.

This shows how much even our wildest fantasies depend on real-life sensations and the use of words as place markers. If we have a rich and varied cultural life, we can build up a wider range of building blocks on which to base our imaginings. The steady reworking of life through art and culture helps unleash our imagination. Modern-day inventions like cartoons help us visualize things that would have been difficult for our ancestors to do just a hundred years ago, but even with our control over language, there is limit to how spectacular or convincing our imaginations can be.

Imagination is limited not just by the richness of our vocabulary or our knowledge of the world around us. Our best attempts at visualization are but a pale imitation of real-life sensations. For instance, imagine a crisp green apple. Those three words should spark a mental image of its shiny skin, the crunch of its bite, the sharpness of its taste. A juicy apple should be one of the easier things in life to picture, but in fact, the image that we manage to conjure up is likely to be rather disjointed and unstable, nothing like the impression we have from looking at a real apple. Bright though our imaginary green might have felt, it would be a poor substitute for real-life greenness and even if we made our imaginary apple as green as possible, the effort of turning up the color "volume" would probably cause us to lose our grip on the shape, smell, or crunchy bite of the apple. Not only would the imaginary apple always feel shadowy compared to a real apple held before our eyes, but trying to strengthen one aspect of our inner picture would lead to the rest of the imaginary qualities fading away. What may have seemed an impressive ability to imagine a lifelike apple in our heads turns out, on closer examination, to be a very pale substitute.

What is true for our imagined apple is also true for all our imaginary creations. Our imagination seems powerfully realistic to us only because we do not examine its flaws too closely. We prefer to believe that our imagined views are as rich and powerful as the real sensations thrust upon us by life itself.

The reason for the weakness of imagination is a matter for speculation. But if we accept the firing-network model, an answer seems obvious when we compare the way the apple net is whipped into existence by real-life sensation and the way it is created through imagination. When we see a real apple, a broad wash of sensation funnels up from our eyes and splashes over the visual cortex in a bottom-up creation of a net of firing cells. Many millions of retina cells join forces to kick the brain into action. But when we imagine an apple, we can use only the thin crust of memories left behind like so much flotsam by the wash of old apple sensations. Our many previous visions of apples will have left a small trace of memory as a pattern connecting a few high-level cells at the peak of processing and our consciously driven imagination has to kick these master cells into life. Once fired into action, the cells should in turn send messages fanning back down the network of nerves established by the original sensation. The farther down the network the messages get, the more cells that are involved, and the more intense our imaginary picture of a green apple. Like a puppet master pulling strings, a few top-level memory cells would reach down to the visual areas and jerk a circular green network of cells into life in an attempt to re-create the experience of seeing a real-life apple.

However, because the lower visual areas are general-purpose and each neuron takes part in many different nets, the original links forged by the apple sensation will be weak—or more likely gone without trace, since the constant pounding of fresh waves of sensation on the main visual surface would have washed it clean of all specific memory traces. So our conscious attempt to re-create the impression of an apple will be able to recruit only a limited number of cells to the effort; the puppet-master memory cells will be able to whip only a small part of the visual zone into action. The resulting image will hardly compare with the freshness and intensity of real vision.

If our imaginary picture of an apple is surprisingly poor, our memory for personal events is probably even more inaccurate and disjointed. We tend to treat our personal memories as if they are faithful video recordings of past events that we can replay in our minds at will, but when we look closely, even a well-remembered scene is in fact a few vague images and snatches of conversation rather badly spliced together.

If we were to recall a chance meeting with someone earlier

in the day—say, a quick chat with a friend whom we had bumped into in a corridor—we would first set the scene, getting the person's face right and the general look of the corridor. Then we would rerun what we could remember of the conversation, and after some rehearsal to bring back as many different details as we could, we might be able to run through a complete repeat of the short meeting. Even then, the re-creation would not be a photographic record of what our senses saw.

For a start, we would probably visualize the whole episode from the viewpoint of a third person—with ourselves as part of the imagined picture—rather than looking through our own eyes, and the length of the meeting would be compressed, with any long pauses in the conversation edited out. Furthermore, the volume and brightness would be strangely unreal. It would not matter if we had been whispering or shouting in the corridor since, mentally, the noise level would seem the same. Likewise, the brightness would have no real dimension. We might imagine a sunlit corridor with warm sunshine bouncing off a wooden floor, but the imagined dazzle would lack photographic reality.

In fact, the more we analyze our memory of the meeting, the more we can see how much it is a practiced re-creation of events rather than a faithful replay of what our senses recorded. We assemble as many elements in working memory as we can and then accept an impressionistic run-through of them as an impressively vivid memory.

Not only is such a memory really a loosely connected assortment of elements, but these elements may themselves be quite fictional. Given how memories are stored, the pattern etched by the morning's meeting would depend very much on how unique the experience was. Its sharp and distinctive features—such as spilling a hot cup of coffee over our hand or relaying a bit of gossip about a colleague—would stick out in our minds but the rest of the experience would blur into a hundred other such meetings. This blur would include the image of ourselves, the other person, the corridor, warm sunlight, the hurried feeling of a busy morning, and any number of other commonplace details.

When we came to relive the memory, we might have dredged up fairly sharp impressions of the details unique to the event such as the spilled coffee and the office gossip. That is, we might have a mental picture of our wet messy hand clutching a plastic cup and go on from that to imagine our attempts to mop up and the brief embarrassed pause that developed in the conversation. Or we might remember what mischief our colleague had got up to, then re-create the words we used to discuss it. However, while these unique memories might have special force and be fairly easily recalled, most of the re-creation would have to be done with general "fill-in" images. We might not be able to recall what our friend had been wearing or the color of the corridor wall, but we would know his face well enough and have a clear enough image of a sunlit corridor to paint in the background detail.

When, therefore, we examine the accuracy of even a well-remembered and recent event, we can see that it is not an accurate video recording of life. Rather, we dredge up the few unique details we can find and then rouse enough general knowledge about such scenes to put together a reasonable simulation of events. Any fragments of the scene that were unclear would be glossed over or guessed at and we would be left feeling that once again we had recovered a lost moment from our past.

It is not surprising that personal memory has to be fleshed out from a few crumbs of apparently hard fact. As we have seen, evolution never meant to provide animals with a faithful video-replay mechanism for their lives. It wanted animals to be able to understand their present, to recognize their surroundings, and to remember the behavior that would be right for the context. Ironically, in many ways a blurred general memory of life is more helpful for achieving this end than the sharp memory for personal experiences that seems so important to humans. As Shereshevskii found with his freak memory, too much detail leads to confusion. Blinded by densely painted fact, we would fail to see the more important broad similarities behind life's experiences. Imagine how impossible the world would be if we had to see a friend from the same angle every time to be able to recognize him.

Evolution thus equipped animals with a memory surface that was quick to weld similar impressions into a blurred general net of knowledge, and when we humans came along with our own reasons for wanting to recall particular moments from our pasts, we had to build on what evolution had provided. We had to learn to become detectives and to re-create experiences from the few scraps of detail and the tangled mess of background knowledge that were all our brains would store.

It might nevertheless be argued that despite the fake viewpoints and all the other flaws, our memories are basically accurate pictures of past events. Indeed, it is true that our memory for personal events is very good—for the broad sweep of what happens. It breaks down, however, when we start going into the sort of minor detail that does not etch a strong trace of its own but becomes woven into a blurred mass of similar memories. When this happens, we have to paint in the background details with educated guesses, but this gets us into a situation where we can no longer tell the dividing line between faithfully accurate memory and helpful imagination—because both are basically the same. The only difference between honest-seeming memory and our flights of fancy is the intention with which we set out. Both are the creation of word-driven habits of thought.

This has been shown in psychological tests to investigate the accuracy of eyewitness evidence given at criminal trials. These demonstrated that the way questions are phrased can have a big influence on the recollections of witnesses. Subjects were shown a film of a mock car crash and then questioned a week later. When the experimenters described the incident during questioning as a smash rather than a collision, the witnesses' estimates of the traveling speed of the cars went up. They were also more likely to report that glass had been shattered in the accident. In a similar experiment, the film of an accident showed a green car passing the crash scene. Most subjects would correctly remember the car as green under later questioning, but when they were asked a series of questions, one of which referred to a blue car going past, the recollections of most people were distorted: When asked a separate set of questions twenty minutes later, with a specific query about the color of the passing car, subjects pictured a blue or bluish-green vehicle. The original trace may have been strong enough for the subjects to remember the color of the car accurately for at least half an hour, but even the indirect suggestion of blueness in later questions was enough to obscure the original trace and change the memory that would be re-created.

Most of us, however, are less concerned about the imperfection of our personal memories than about how much of life we routinely forget. Rereading an old diary or visiting a childhood town can make us realize how much of our lives is lost to us forever, buried under the blur of a lifetime's experiences. Without diaries or snapshots to prod our memories, we might die without ever again stumbling on the pleasure of a fishing trip one long-ago summer or of a watch given for a ninth-birthday present.

We should finish our discussion of memory by looking at what is often felt to be a second, separate form of long-term memory—factual memory, the schoolbook-type memory for facts, figures, and general knowledge. Psychologists normally classify factual memory as quite different from personal memory, but factual memories are just like personal memories in the way that nets are laid down by washes of sensation and later recalled to glowing life in the conscious plane.

For example, most people will have a net of general knowledge about Joan of Arc or South American condors, built up during the course of their lives from hearing dozens of references to Joan of Arc, reading accounts of Peruvian wildlife, watching nature documentaries, or seeing pictures of Joan of Arc at the stake. All this reading, listening, and watching is as much a part of our experience of life as the time spent on such personal things as drinking coffee or eating picnics, and memory traces will gradually be built up in just the same way. At first each snippet of knowledge of Joan of Arc or condors will leave a unique imprint, but with time we will build up the same sort of rich blurred nets of knowledge as for things like picnics or coffee drinking. These nets of knowledge will lie dormant in the gray matter until the tug of recognition calls the buried information to life again, when the nets are likely to be displayed in the conscious plane in a variety of ways. With condors, for example, we may see mental pictures of big black birds soaring over the Andes or hear the harsh cries remembered from a TV nature series; we may hear ourselves mouthing facts about the width of the wingspan or the rarity of the species. This is important since it shows how we only "know" something when we are replaying it in the sensory processing areas; for the rest of the time a knowledge net about condors is like an unread book on a dusty library shelf.

Both personal memory and factual memory are, then, based on the same brain mechanisms. Both are the result of sensations that have splashed through consciousness, leaving a memory trace in their wake. This trace is either so powerful that it leaves a unique imprint or it becomes woven into a blurred mass of similar-shaped memory patterns. From these buried memory traces, we can try to call the original sensation back to life. Through the skillful use of words, we can trigger the right networks of nerves and re-create nets very similar to the ones that fired before. This leads to sensations apparently reentering our consciousness, and, depending on what sort of memory we are trying to re-create, we will feel that we have either called to life a lost moment from our past or brought to mind a fact learned about life.

But while factual memory uses exactly the same mechanisms as personal memory, there is an important difference. Personal memories are mostly constructed by ourselves without any help or correction from those around us, whereas factual memory is knowledge packaged by the societies we live in. Factual memory is the distilled wisdom of a civilization, parceled up for easy learning and taught in systematic fashion using every means from proverbs to textbooks and churches to classrooms. Personal memories are formed almost by accident but factual memory is deliberately stamped into the minds of impressionable youngsters because it is usually considered vital to the workings of society.

This look at memory should have shown that it is not the simple, automatic mechanism that we commonly take it to be. Our brains, with their fine meshwork of billions upon billions of nerve connections have a built-in ability to recognize and associate but it takes the skilled use of language, an inner dialogue, to jog buried personal memories and imaginings into life.

This must mean that until our hominid ancestors developed language, they would have been as much creatures of the present as the rest of the animal kingdom. Until man developed the habit of searching his own mind, he would have lived just for the moment, reacting to the shifting patterns of the world around him. He could not have browsed through childhood memories or daydreamed about the future in his idle moments. But once he learned the trick of controlled recall, he could create an inner world of memories. Humans could start making plans, building up personal life histories—and even begin to be self-conscious.

Self-consciousness is clearly based on this artificial memory control that came with language. Self-awareness is properly defined as the ability to see and remember what goes on in our own minds—the ability to think about our own thoughts, relive old experiences, and recapture old feelings. We know who we are only because we have a stream of memories stretching back to our early childhood which can be recalled at will—or at least after a determined search. Control over our memory banks allows us to form the sense of personal identity that is the core of our feeling of self-awareness. Literally, we are self-conscious because we can remember the fact that we are conscious. Animals are, simply, conscious—which, as we shall see, is quite a different thing. Their brains may buzz with an intelligent awareness of the world outside but they have no mechanisms for looking inward and observing this process of consciousness at work.

Memory control is essential not only for a broad understanding of the fact of our own existence but also for a moment-to-moment awareness of the activity in our heads. It might be assumed that the brain is automatically aware of everything that takes place inside it, as it is happening, but if we look closely, we will see that this is not the case. The reason is that the brain is too close to itself. Just as the eye can see, yet cannot see itself, the brain can be aware without necessarily being aware of itself. To be self-aware, the brain, like the eye, would have to get outside itself and view its activity from a physically separate vantage point.

Clearly, the mind is in no position to step outside the body to reach this new viewpoint. But there is a second possibility: If the brain cannot step back in space to observe itself in action, then, through the control of memory, it can take a step back in time. Once humans gained memory control, we could think something and then replay it to examine what we had just thought. We could trap fleeting impressions after they happened and so savor all our thoughts, sensations, and feelings a split second after they had passed through our minds. They would not all be swept away into history by the constant wash of life's sensations as they would in an animal's mind. Memories of our mental states could be recalled at leisure, giving us the ability to focus on what existed in our minds and so become truly self-conscious rather than merely conscious. Humans learned to step back from their minds in time in a way that they could not do in space, and so were able to become self-aware.

Although it may seem odd to claim that we cannot both think something and also be self-consciously aware of what we are thinking, this is because we usually focus only on slow-moving mental events and rarely notice the separate acts of first doing and then self-consciously reviewing what we have done. The split is more obvious with physical actions, like kicking a ball or swinging a bat, that take place very rapidly. When we strike a ball, we are aware of what we are doing but only in the pure sense of seeing a ball and feeling the way our body acts. If we want to savor or criticize our swipe at a tennis ball, for example, we have to grab the memory quickly and run through an action replay. The difference between the first time, when we just swung at the ball, and the second time, when we were self-consciously aware of the action, is clearer because there is a distinct time lag. Sportsmen often describe their best shots as instinctive—they do not think about it, they just do it. Although the time lag that is needed to be self-consciously aware is most obvious with fast physical actions, the same principle applies to all the mind's actions.

Having looked at memory, we shall next explore another special feature of the human mind—thought. In many ways, thought is another way of looking at memory—simply taking memory one stage further and animating it. Also, like memory, thought has natural foundations common to all animals but given a new twist by humans through the use of language.

FOUR

Thinking Aloud

Thought is often spoken of as the slow step-by-step chains of logic and words that we build to reach weighty decisions. However, this is only the most overblown and artificial form of human thought—a style of thinking that sees little everyday use. A better and wider definition of thought is that it is whatever happens to be running through the tight central focus of our minds at a certain moment.

When we look closely, we see that our minds are usually filled with a tumbling succession of thought fragments. We have flashes of memory, bursts of words, fleeting gasps of emotion. Bits of thought seem to zap around in our heads like flies in a shuttered room, zooming in and out of focus. Just reading that last sentence should have provoked a brief flurry of thoughts: perhaps a mental image of a hot dusty room with drifting net curtains and half a dozen wandering bluebottles, or maybe a half-formed, unanswered inner question such as: "My thoughts come out of somewhere . . . but where?" These thoughtful diversions from the printed page would have taken only a split second. Moreover, our reading may have been interrupted by a constant stream of niggling little distractions, for instance, the distant roar of a jet might briefly have caught our attention or we might have noticed a dull ache in our back, the swoop of a bird past the window, an itch on the back of the leg, a stomach rumble of hunger, or the sudden barking of a dog. Yet despite this tremendous amount of mental activity, if someone asked us what we were doing, we would unhesitatingly reply that we were reading and we would feel that we were doing so with almost tunnel-vision intensity.

The way flashes of thought and sensation tumble so quickly through our awareness makes it hard for us to get a firm hold on what is happening in our minds when asked to describe it. Even when we try to give an honest answer to ourselves, we tend to overlook all the petty distractions and remember only the main thrust of what we thought we were doing. But the mind—like the nets that form it—is defined by what it is doing at an instant. If we ignore the smallest distraction that catches our attention for even a moment—such as the pause to identify the roar of a passing jet—then we misunderstand the true nature of our minds.

It is, however, very difficult to take every fleeting thought and sensation into account when examining how our minds work. We can ask ourselves questions about what we were thinking just a moment before, but such questions pile up more rapidly than they can be answered. Too much happens too quickly to recover every little detail. Even worse, we cannot think about an earlier thought without creating another, so that we are always left helplessly one step behind as we chase after events in our own speeding minds.

This shows how limited is our ability to really follow what

goes on in our own minds. As shown earlier, to be self-aware we need first to do something and then be able to reflect on it an instant later. We cannot step outside our bodies to observe ourselves, so we have to create a metaphorical distance by stepping back in time. This metaphorical distance gives us rather a limited view of ourselves, so normally we notice only the most obvious things happening inside our heads. We do not have time to stop and examine in detail every fragment of thought and instead gloss over this background mental activity so that if we ask ourselves about it, we find we lack the detailed knowledge to answer such questions. This is when we may feel that the mind is somehow mysterious and other-worldly, but in fact we have simply not trained ourselves to watch our own thought processes as well as we could.

Coming back to thought, we feel that our minds follow a clean-cut thread of ideas from premise to rational conclusion, ignoring all the stops and starts, all the distractions and all the stray ideas that clutter up our minds. Nevertheless, chains of consciously driven thought are an important part of what goes on in our minds and we will now take a closer look at how we assemble these chains and make some headway against the buffeting from life's steady stream of distractions.

Obviously, thoughts move along one step at a time. One instant leads to another and simple flashes of thought become linked in a long chain just as single frames in a filmstrip create a moving picture when run through the projector. The puzzle is what lines up frames of thought in the right order and then propels them through our consciousness.

The problem with answering this question is that we tend to picture thoughts as if they rattle around in the head like peas in a pod or, as said earlier, flies in a shuttered room. All we have to do is pluck a thought out of the air to have a complete answer to a question clutched in our hand. But this is the wrong sort of analogy. Thoughts flash into awareness as nets come alive against the gray background of the brain. As has been seen, the pushing and prodding of attention can inflame and spread the net so that nerves on the outskirts are tugged awake. Since each net is connected by habit to many other related nets, the awakening of one will send flashing signals down the linking nerve fibers to rouse others in the chain. Then, as the first net of nerves dies down or the nerves at the center become numb, the newly alerted areas take over as the nets in the limelight of our consciousness.

In humans, what typically happens is that we spot something in the world around us that sparks a chain of thought. The chain may be brief or long. When we hear the passing roar of a jet, for example, we may get no further than identifying the source of the noise or, having hit on the memory net with information about jets, we may follow an endless trail of associations, leading us to such memories as trips abroad or the blandness of airline food.

In our earlier example of spotting a picnic while driving along on a summer's day, initially an odd grouping of people prompted a puzzled feeling and after a second or so of uncertain fluttering of possible memory nets, arousing half-pictured images of religious ceremonies and open-air experiments, the right chord of recognition was struck, triggering a jumble of memories about picnics. But the initial difficulty might have set us off on a fragmented inner dialogue while pursuing the chain of thought, on the lines of: "picnic! . . . of course . . . that was a hamper . . . weren't there kids? . . . one had a hat . . . a paper party hat? . . . perhaps they had a tartan rug like . . . like they have in movies." By this time the original picnic scene would be fading into the gray background and recollections of old movie scenes would perhaps be clamoring for attention at the fringes of our awareness—

until a sharp bend in the road happened to drag our full attention back to the task of getting home in one piece.

As these examples illustrate, the natural mechanisms powering thought processess work like the spreading of stains or the toppling of dominoes. The paths that thoughts will take are laid down by experience but stay hidden in the shadows until illuminated by a wave of nerve firings. Each chain of thought is thus not a free-moving body, floating through mental space, but a stepping-stone sequence of nets being lit up in turn. Each step is firmly rooted in the surrounding gray network as if atop an unseen pyramid of supporting experience. The tip may be all that we are aware of at a given instant, but the pyramid is what gives the process of thought its real weight.

We hate to think that our minds are not free, and we believe that our thoughts can fly where they like and that our minds are our own. But in fact, the way our thoughts flow is heavily dictated by experience and the cultures we live in. We have already seen how imagination is reined in by our inability to imagine anything that is not in some way based on real-life experience. In much the same way, we will also see how much our cultures dominate the direction our thoughts take. However, the freedom of our minds is really a nonissue, for the richness of modern culture gives us more freedom than most of us know how to handle. Even environment-led animals such as chimps and dogs look fairly independent-minded to our eye. It is only when we look at simple animals that behavior appears indeed limited and reflex-driven.

Very simple animals, like worms and jellyfish, can hardly be said to have any thoughts at all. Thought has meaning for an animal only when it can at least hold some internal representation of the world in its head, making the sort of mental maps that lead to consciously felt nets. So worms and jellyfish do not really think. Their nervous systems simply react and adjust to the world in hard-wired fashion. Moving on to fish and reptiles, we see signs of intelligent minds at work, although their cold-blooded metabolisms put a limit on how much they can invest in energetic brain work—or how much they could do about tricky situations if they had the wit to realize they existed. However, experimenters have found enough examples—like lizards that are able to find their way through a maze to a warm lamp—to convince them that even reptiles have some internal picture of the world.

The higher up the animal intelligence scale we go, the more knowledge an animal can bring to bear on any particular situation. A pet frog will never learn to recognize its owner. It may have the brain to spot a hand looming down to grab it out of its tank but it will never be able to distinguish between a friendly hand and a potentially dangerous hand. Nevertheless, knowing that a hand is coming down is no mean feat, meaning that the frog has at least enough of an inner representation to distinguish between a falling leaf and a descending palm. With pet dogs, the amount of information supporting any course of action is sizable. A dog will recognize its owner's hand and probably have a fair idea of the mood of the owner, depending on whether the dog has just retrieved a stick or chewed up a shoe. In the dog's world, previous experience and learning play a big part.

With dolphins and chimps, we see the creative leaps in thought that so impress us in humans. Dolphins quite often mimic the swimming actions of other dolphins and even other sea animals like penguins—a trick that clearly requires insight. One young dolphin reportedly saw its human trainer drag on a cigarette and blow out a cloud of smoke, after which the dolphin dived to its mother, took a mouthful of milk, and imitated what it had just seen.

In a more formal test of reasoning, young chimps were left in a cage with a long stick and with a banana just out of arm's reach on the other side of the bars. The chimps had never had a chance to play with sticks before and did not attempt to use the one in their cage to hook the food nearer. They tried stretching through the bars to reach the banana a few times and then gave up. However, another set of chimps, who had been given three days to play around with a stick before the test, all saw the answer within twenty seconds. Their apparently aimless play had allowed them to build up a rich net of information about sticks.

Solving the problem may have gone something like this. Attempts to reach the banana would mobilize what we might call a general "banana reaching" net. Recognition of the stick would wake up the knowledge about sticks. Then, with what humans might call a sudden flash of insight, the two aroused nets would have linked up and the chimp would have suddenly "understood" that its stick-handling knowledge could solve the problem. Humans at this stage would mentally picture themselves reaching out with a stick and immediately get an "aha" feeling of satisfied recognition, telling them that they had hit on the answer. Perhaps in some wordless fashion, chimps too get a mental picture of the solution.

This building of bridges between aroused nets to answer problems is what we might call natural thought. We build up knowledge of the world and then recognize situations where that knowledge can be applied. Humans use this natural thought all the time. In an experiment similar to the banana problem, children aged three to five years were given the task of clamping two sticks together to reach objects on a tray at the other end of the table. Children who had first been allowed to play with the sticks and clamps were not only quicker at solving the problem than those without such experience but also quicker than another group who had actually had the right method demonstrated to them by an adult. Making a rake appeared to depend more on a rich general net of understanding than a quick instruction session.

Our big-brained hominid ancestors must have had considerable raw reasoning power even before speech came along. However, language made a huge difference in enabling humans to take inner control over the direction of thought. It made it possible for humans to break free of the tyranny of the present and chase thoughts far removed from the events of the moment.

We have seen how animal thoughts are tied to the present. It is said that a dog can be afraid its master will beat it but not that its master might beat it tomorrow. For the dog, out of sight is out of mind, and unless the owner is present, there is no stimulus to awaken memory nets about possible beatings. It is almost as if evolution built a vehicle without a driver. The brain is a powerful engine and the body provides a set of wheels but the vehicle is driven by the demands of the environment. A cat sees a mouse and pounces but when the cat sits in the sun with no mice to be seen, it is not, as has been argued, likely to be daydreaming about extra-plump mice or reliving earlier mouse-hunting expeditions.

True, this picture of a blank existence is complicated by the stirrings of emotions. Emotions will be looked at in more detail in a later chapter, but they could be broadly defined as the sensations that flow from the many sense organs dedicated to monitoring the state of our own bodies. Just as we need sense organs to build up a consciously experienced picture of the outside world, so we need sensations, such as hunger, pain, and thirst, to tell us what we should be doing to satisfy the demands of our bodies. As the decision-making center, the brain has to have a window looking on to the inside world of the body, as well as a view of the world outside, if it is to act intelligently.

This means that an emotion can stir an animal into life quite independently of what is going on in the outside world. A sudden pang of hunger rumbling up from the guts could make an idling cat jump up and go in search of its dinner bowl; such feelings are as much a prod from "outside" the brain as the squeaking of a mouse. So even when being prompted by an emotion, an animal's mind is still tied to the present. It is not reacting to inner chains of thought, but responding to the push and pull of events around it.

Humans, on the other hand, have constructed their own inner driver. Language gives us access to memories of the past and fantasies about the future. We can break away from the grip of the present and think exciting thoughts about the meaning of life or what we should buy at the supermarket this afternoon. Once humans made the fundamental break with the present tense through memory control, it was possible for a spectacular blossoming of human thought to take place. However, we should not forget that we actually still live our lives mostly in the present. We are usually reacting to outside events and the urgings of our body rather than sitting around contemplating our navels. But the possibility is always there for our minds to go off at a tangent to the outside world. And even when the environment is firmly behind the driving wheel, our internal driver is usually in the background, shouting helpful comments from the backseat.

Returning to the natural basis of thought: A typical chain of animal thought goes through the stages of sensation, recognition, knowledge mobilization, and behavior. The problems are always posed by the present, and how intelligently an animal reacts will depend on the amount of knowledge it can bring to bear. Chimps could use a stick as a rake only after they had built up a net of general knowledge about sticks. With more experience, some chimps learned how to jam two pieces of bamboo together to make a very long stick to reach the banana. In the wild, chimps have learned how to make their own sticks for termite gathering by stripping the leaves off a twig of just the right length and slenderness. With learning, the net about stick use can become very rich, spreading and growing like a stain in the cortex. Many individual memories gradually become blurred together so the chimp is left with a large all-purpose net.

Humans are convinced of the importance of having a clear memory for all the events of their lives, so we assume that evolution's ultimate aim would be to provide animals with a memory so good that they could vividly remember the first time they waved a stick around. But evolution is more interested in providing the sort of blurred mass of memories that can drive an animal's thoughts along in a wide variety of situations. Chimps have no special need to replay mentally their original experiences with sticks, whereas a rich net of stick-using knowledge will equip them with the understanding to tackle problems cropping up in the real world. We can therefore see why the brain forms the sort of memory nets it does, nets that are tuned to soaking up a mass of details rather than making a photographic record of life. And they provide a natural method of thinking: The awakening of a net will bring to mind the sort of knowledge needed to solve the problems of the moment.

Humans may have developed language and discovered new ways to use the brain's memory surfaces, but we still rely heavily on the broad understanding of life that natural memory gives us. This is the sort of knowledge about how branches bend and rocks fall that forms the internal backdrop to our thinking. As we have seen, this inner backdrop of common-sense knowledge is so basic that we take it for granted and forget it has to be learned before we can go on to more advanced problem solving. For example, babies need to learn that things still exist even when they disappear from sight—as when a mother hides a rattle behind her back—or that there is one space that contains everything in the world, including the baby itself, rather than a multitude of separate spaces such as the world of things inside the mouth and the

worlds touched by the hands, seen by the eyes, or heard by
the ears.

These sorts of facts are learned by the time a child is about
one, but even at four or five years of age, children often have
trouble with seemingly obvious mental tasks. If all the water
is poured out of a tall thin glass into a short fat glass, the
child thinks that there must be less water in the fat glass
because of its lower level. Common sense does not yet tell it
that if no water has been spilled, the quantities must be exactly the same.

There are many other examples of how much every child
has to learn before it can start on the more artificial habits
of human thinking, such as deductive reasoning and scientific method. These types of thought do not appear until a
child is about twelve years old—and even then they need to
be actively taught rather than developing as a matter of course.
Also, a look at these grander types of thought soon shows
how dependent they are on the solid foundation of common-
sense reasoning.

It is a great myth of modern man that we are inherently
rational. We talk about deductive logic as if it were wired
into our brains and as if it were the main difference between
us and other animals, but formal logic is a very recent creation of modern man. The artificiality of Western logic is
highlighted by the navigation feats of the Truk Islanders in
the South Pacific, who regularly sail hundreds of miles between little coral islands, finding their way by "feel." The
trained Western navigator would find his way around by charts
and measurements. If asked at any time where he was, he
could point to the map and give a logical step-by-step account of the course he must follow to get to his destination.
The Truk Islander, however, has a mental picture of where
the island lies over the horizon and points his boat in the
right direction until he gets there, keeping an eye on the waves

and the winds and the general look of the sun and stars. Without any conscious calculation, he can keep a feel of where he should be heading even while continually tacking from side to side when sailing into the wind.

The Truk style of thinking is seen as primitive because the islanders are unable to articulate the rules by which they are maintaining their course. When asked, they shrug and point to where they think the island lies. Early humans must have existed for hundreds of thousands of years with such natural methods of thought, doing what they felt was right from experience or custom. Today we would describe such thinking as intuitive or instinctive, but it is really just allowing our rich nets of knowledge to surface in the conscious plane and then taking heed of the images they present. Modern man has been so trained to trust only thoughts driven along by logical chains of words that this inner knowledge is treated nervously. We may sometimes act on hunches and gut feelings, but we are happier when we can talk of the logical reasons for doing something. Indeed, we will often find an intuitive answer and then look for logical reasons to back up our decision.

Logical thought is not of course wrong. It is a tremendously powerful tool that has brought us the technological revolution of the last five thousand years. It is, however, of far more limited use than commonly supposed and can be dangerously misleading unless firmly rooted in natural net-driven thought. Before looking at the dangers, let's be clear about the job of logical or scientific thought. Basically, such thought strips away all the obscuring detail of life so that we can see the bare-bones story of how things work. Once we have boiled the world down to its essentials, we can see the relationships or rules that govern life and use these newly discovered rules to solve fresh problems.

Mathematics is the most extreme science for stripping nature down to its bare bones. While biology or chemistry still talks about physical objects such as cells or elements, mathematics deals only in symbols. It aims to reach a pure description of the relationship between objects without any mention of what the objects themselves might be. So $2 + 2 = 4$ no matter whether we are talking about two elephants or two acorns. By stripping away tusks and tails, nuts and stalks, we can discover the underlying order of the world and put it into a language of symbols.

Math has many branches because it focuses on different sorts of relationships. Arithmetic is a system of rules for describing relationships between objects, geometry is a system for describing the shape of objects, and logic is used to describe what is already known about objects in useful new ways without breaking the underlying relationships. Mathematicians are gradually discovering new tools—such as algebra, calculus, and statistics—that capture more complex relationships. But while math is a powerful tool, it is good at describing only the simplest sorts of relationships. It is stretched when it gets much beyond the relatively clear-cut worlds of physics and chemistry.

Despite the limitations of rational thinking, we value it because it gives us a framework with which to turn thoughts into public property. We can state the assumptions and logical steps that go into our arguments, which can then be shared and examined by those around us. Truk Islanders may be able to reach coral atolls hidden over the horizon, but they cannot tell other people how, whereas Western navigators have the geometry of charts and the logic of course setting to describe the underlying thoughts in very precise form.

On close inspection, the rational thinking that is supposed to be the hallmark of modern man is really a rather specialized skill. It is extremely useful for the sort of publicly discussed thought that underpins science, but if we analyze what takes place in our own minds, we will see that even our most abstract thoughts are not dry strings of symbols. In the privacy of our heads we tackle problems with metaphors drawn from everyday life rather than by following computerlike chains of logic.

A good example of everyday problem solving is the puzzle of how to treat stomach cancer. A tumor can be killed if hit for long enough with a beam of radiation—but only at the expense of also killing the layers of good tissue that the beam passes through. In a psychological test, few subjects spontaneously came up with the right answer, but when they were first told a story about an army attacking a heavily defended castle from many different directions or asked to imagine the spokes of a cartwheel, most of them promptly saw the solution: that the radiation beam should be moved about to hit the tumor from many angles without staying in one place long enough to damage the flesh in between. The mental image of a defended castle or spoked wheel had provided a working model that could be matched against the outlines of the problem. In a flash of recognition, they saw that the metaphor fitted the bill and the solution became obvious.

In terms of nets, we would first assemble the elements of the question about treating stomach tumors in what might be called the problem space of working memory and let the words spark an image of the situation. For example, we might visualize a patient on an operating table with a big ray gun pointing at his stomach. We would then hope that this vivid image would strike a chord somewhere in our memory banks. If nothing came immediately, we might play around with the image—perhaps mentally wobble the ray gun. This could be just enough for us suddenly to make the big connection with our recently aroused memories of sieges and wheels, or rather, the natural recognition ability of our brain surfaces would react to the resemblance between our imagined picture of the operating table and the still warm nets on sieges and wheels. Once the match had been made, the metaphors would be triggered brightly back into life. Then, with both the problem image and the answer image firing in our consciousness, we would see we had the solution. After that would come the relatively easy task of putting the answer into words. We would tell the psychologist doing the test that all we needed to do was attack the tumor from many angles and then sit back pleased by our logical minds.

The point of this example is that being rational is really something that we do in public once we have cracked a problem in private. And we solve problems not with dry logic but by hitting on the right mental images. This metaphor-driven thought is in turn based on the sort of rich blurred nets of knowledge that all animals use to power thought. The difference is that humans have control over the process. We do not need both the question and the answer staring us in the face, like the chimp in the banana and stick experiment. We can imagine problems in the privacy of our heads and then scour our memory banks for possible matches with other images.

The richer our net of knowledge about something, the better the metaphor it will make, which is why we use everyday objects to mimic the way unfamiliar or even totally abstract things are going to work. When scientists talk about electrons, for instance, they think about them in terms of waves or little balls. The electrons are not in fact much like either, but because they are outside what we can directly sense, we have little choice but to build a secondhand picture from something familiar, such as balls and waves, so that we feel we have some understanding of the world of the electron.

Looking at what we consider to be abstract or rational thought, it becomes clear that we do not really feel we understand something until we ground it in commonplace experience. A list of all the known properties of an electron remains dry words until we take a rich knowledge net about ricocheting balls or rippling waves to animate them. The same applies whatever the subject. The whole of human intellectual achievement rides along on the back of metaphors.

In net terms, every word we use must sit atop a net of knowledge, striking some chord with our memory surface to have any meaning. Everyday words such as *dog* or *apple* lead to rich stores of knowledge, for we have seen, heard, and smelled thousands of dogs and apples, so we have a vivid idea of the sort of things they will do. This knowledge can be used to push our thoughts along whenever we bring dogs and apples to mind. Even verbs and adjectives sit atop vast stores of knowledge. We know from experience what *red* or what *running* is and the many real-life forms each can take. But the nets underlying such abstract terms as electrons or justice are mostly dry textbook learning—what we remember reading or being told in the classroom. We might be able to regurgitate this learning parrot fashion, but we would not have the richly detailed memory nets that go with words about everyday experiences, so to flesh out abstract words we use metaphors. Our dry net of knowledge about justice, for example, is animated by the moral tales and fables that we are told as children, which build up a mental picture of situations that demonstrate fair play. And just as we use the metaphor of a ball for an electron, justice is portrayed as a blindfolded woman holding a pair of scales.

This raises the point that the depth of human thought must be limited by the metaphors available. The natural world gives us many images, such as flowing water, burning fire, bending branches, and so on, but these can arm us only with a restricted number of mental metaphors. However, when man started up the path of technology, he also created a wealth of new metaphors, and down the years, the dominant metaphors of cultures have reflected their latest technologies. The ancient Greeks used images of pottery and spinning to help them understand the heavens and the stars: Plato spoke of the universe as a spinning spindle; hanging balls of yarn were the celestial spheres with gods measuring and cutting the thread of fate. By the Middle Ages, clockwork was the most popular motif for explaining the world, and when the Industrial Revolution brought steam power and hydraulic systems, these became the new driving images for scientists.

Freud, for example, argued that if psychological symptoms were repressed, then—as when water is forced down by a piston—the personality would simply spring leaks elsewhere. This hydraulic metaphor was convincing for many years, until a different school of psychologists came along, who did not believe neuroses and sexual feelings welled up like an explosive head of steam in a boiler but thought they were learned behaviors that grew on the surface of a personality and could simply be pruned back without great harm. This new analogy led to advances in treatment whereby patients unlearned behaviors rather than going through exhaustive analysis to tap their "subconscious" roots.

Today it is computers that dominate as metaphors, with psychologists talking of the programs and operating systems of the human psyche. Computers have only a passing similarity to human minds, but as the most complicated technology humans have created, they provide some of the best metaphors we have come up with so far. We can, however, get carried away by metaphors. If we talk about the mind as a computer, we may become blind to the many ways it is not like a computer. A rich mix of metaphors seems to give the broadest possible feel for a subject. In this book we have talked about nets, pyramids, fragmented maps, foundations, extensions, spotlights, stages, and dozens of other things to try to illustrate how the mind works.

The very fact that we need to struggle so hard to understand our minds should tell us a lot about self-consciousness. Self-consciousness is very limited. Language gave us the space to step back and see ourselves, but we have no automatic ability to understand what we see. We need good metaphors to feel we understand ourselves. Our cultures arm us with some ideas to start us off but there is a lot more that could be done to raise our self-awareness.

This ramble through the subject of human thought should have helped make a few points clear. Man's hominid ancestors of a couple of million years ago would have had powerful, if wordless, thought. Like chimps, they would have had the wit to make tools and absorb a global feel for the way the world behaved, but they would also have been creatures of the present, driven by their environment rather than guided by inner thoughts. Early man's large brain would have given him tremendous natural intelligence but all his gathered knowledge about life would lie dormant in his memory banks until triggered by some event in the outside world. Having once been roused, these nets would quickly fade from conscious awareness, disappearing back into the gray background. The hominids needed language to take the quantum leap forward and break free of the straitjacket of the present tense.

We modern humans have, however, become overly impressed by our capacity for abstract thought. The ability enables us to create useful formulas to work with, but abstract thoughts must resonate against a rich background of imagery that only real-life experience can provide; otherwise we do not feel we fully understand the thoughts being expressed. They remain a dry, colorless string of words rather than springing to vivid life inside our minds. As we complete our look at nets with a few thoughts about thought, it should be clear that all the mind's facets are really different ways of looking at one underlying neural surface. Learning, memory, imagination, thought, perception, and awareness are all different ways of viewing the net-forming properties of the brain. It should also be clear that the human mind differs from the raw animal mind in having picked up extra habits of thought. The animal mind is made up of perception, recognition, awareness, simple thought association, and environment-led attention. The human mind has perception, memory and imagination, complex habits of thought, inner-driven attention, and self-awareness. These extensions to the animal mind have been built out of language—although exactly how that happened has not yet been fully examined. First, however, we should look at how humans learned language, which means returning to the evolutionary story of our hominid ancestors, whom we left shivering in the ice ages a couple of chapters back.

FIVE

From Baby Talk to Strong Language

Language was almost certainly the key to the dramatic rise of our direct ancestors, *Homo sapiens*, who spread out from their original Middle Eastern or North African homeland about forty thousand years ago to colonize the globe and replace all earlier relatives. But deciding how or when language might have emerged is a difficult task in the absence of concrete evidence, and we can only search for indirect clues and hazard a few educated guesses.

Such speculation has a bad name among scientists. Indeed, at the turn of the century, the Linguistic Society of France had a standing ban on papers about the origin of speech because of some of the nonsense being presented. One idea—known as the Bowwow theory—suggested that early man started by imitating the natural sounds of the world—going "quack" or "cuckoo" to alert his hunting partner to a gaggle of ducks or whatever. The so-called Yo-Heave-Ho theory, on the other hand, saw language developing from the sort of noises made by a group of early men involved in some communal effort. Scholars imagined the group all grunting to roll over a boulder or shift a tree trunk, coordinating their efforts with the Stone Age equivalent of "one-two-three-heave!"

Other possibilities include the theory that speech developed from an ancient gesture language or that it started off as a baby's cries for the attention of its mother. Some speculated that language originated as a form of sexual display, like a peacock's gaudy tail, a theory that supposedly gave a new meaning to the phrase "whispering sweet nothings" and which drew support from the way the male's voice breaks at puberty.

None of these ideas sounds very convincing. A major difficulty is explaining how the voice box and brain of early man could have evolved to the point where the first stage of speech became physically possible. Evolution is ruthlessly economical, and it would not take early man halfway toward speech unless the halfway stage had some real usefulness. To find a better answer, we will take a closer look at the communication abilities of man's closest living relatives, the great apes.

Surprisingly, given the importance of noises to man, the great apes are fairly silent animals, tending to shuffle around the forests in quiet bands. By contrast, our less intelligent monkey relatives are noisy creatures, chattering and shrieking much of the time. They have developed quite an advanced use of sounds to communicate. The vervet monkey of the African savanna, for example, uses different alarm cries to signal different types of danger: a leopard sighting prompts a bark, sending the troop scurrying for the trees; a special snake alarm, a chuttering sort of noise, has all the monkeys standing on tiptoe to peer down into the long grass around them; an aerial threat, like a hovering eagle, is greeted with a rasping screech that sends the troop tearing down out of the exposed branches of a tree. Each threat provokes the call that triggers the most suitable reaction.

Apes, on the other hand, do not depend as much on calls and cries to keep the group acting in harmony. The orangutan of the Asian jungle lives a fairly solitary life, not requiring such calls, while the slow-paced life of gorillas does not perhaps need cries to coordinate the action of the band. The chimpanzee is the noisiest ape, yet it still uses only about a dozen different noises—grunts, hoots, screeches, and whimpers—compared to the hundreds of different sounds the human vocal organs can produce. However, vocal sounds are not the only way in which apes communicate. Chimps employ a rich variety of gestures and facial expressions to keep in touch with each other, and more important, there is an intelligence behind the exchanges that makes for a level of understanding unseen elsewhere in the animal world.

To appreciate this, we need to distinguish between different levels of communication. The simplest is the alarm call—the instinctive or wired-in reaction to danger that automatically alerts the whole group. Vervet monkeys, as mentioned, make sophisticated use of such calls, but most flock and group-living species have evolved some sort of wired-in alarm call. Although often appearing to be impressive examples of communication, alarm calls are both automatically triggered and automatically understood, as is obvious by the way seagulls and other birds squawk their alarm cries even when by themselves. And while the vervet monkey may appear to be intelligent with its variety of calls, this simply shows it has a strong evolutionary need for different sorts of reactions to different threats.

Birds make a lot of use of songs and cries. As well as alarm calls, birds have mating cries, territorial singing, threat displays, and food-begging chirrups. These noises are all genetically programmed and require little thought, so they do not qualify as deliberate attempts to communicate in the human sense.

A higher level of communication than wired-in instinctive calls is the expressive or emotional cry, where an animal gives vent to its inner feelings. A chimp, for example, might hoot with anger or screech with fear. Such a noise could be said to be genetically programmed like a call, since a chimp does not have to learn to screech or hoot and has quite standard responses to its feelings. The difference is that an emotional cry does not trigger a guaranteed response in the listener, who needs a certain intelligence to interpret the reason for the unhappy noises and to react appropriately.

For example, if a young male chimp provokes an angry hooting in a dominant male by becoming too interested in a female in heat, he has to make sense of the threat. From experience, the youngster should know that the dominant male will always protect his right to have first access to a female and that he risks a nasty fight. This sort of reading of social situations takes considerable insight on the part of the listener, but only a limited communication ability is required by the dominant male, who is simply giving vent to annoyance at what he sees. The angry noises are the first stages of a process that will escalate into a full-scale attack if the youngster continues its advances.

Calls and cries are effective but they are not what we would describe as true forms of communication, where an animal deliberately sends a message to another member of its group rather than just giving voice to an emotion—where signaling comes under the control of the conscious cortex rather than the subconscious emotional system. Chimps can indeed communicate in this deliberate fashion. The young male chimp just mentioned, after his setback, quietly gestured to the female to follow him to a place out of sight of the dominant male: Sitting several feet away and glancing at her out of the corner of his eye, he extended his hand a couple of times in a begging gesture, stamping his foot to make sure that he had her attention, then slowly wandered off with one eye over his shoulder to make sure she was following.

This sort of communication ability is what makes chimps appear far more socially advanced than any other animal. They may have a simple repertoire of noises and body language, but the intelligence with which these signals are used and interpreted makes a big difference. Only recently has it been realized how well chimpanzees can communicate. Most of the observations have come from a troop of wild chimps at the Gombe Stream Reserve on the shores of Lake Tanganyika and from a captive group in Holland's Arnhem Zoo.

In any social animal, the most basic need is to establish a group hierarchy so that individuals know their place and the animals can live peacefully together. With birds and most mammals there usually develops a strict pecking order, with each individual finding its own rung on the social ladder. Chimp troops have a dominant male who generally has first rights to food or females and is responsible for breaking up squabbles among lower-ranked individuals. But because one male is usually not strong enough to rule alone, he has to form an alliance with another male to control the group and has to be prepared to grant certain rights to his lieutenant. Such alliances can last for years and would not be maintained without a strong scaffolding of social communication.

At Arnhem Zoo, when a maturing male, Dandy, started to challenge for leadership of the troop, his tactic was to tag along with the dominant male, Nikkie, and his partner, Jerome, while the two went on their regular patrols of the zoo's open-air cage. Dandy appeared to be trying to disrupt the ruling alliance by boldly walking between Nikkie and Jerome as if he were himself a top male. Dandy also tried to take on some of the dominant male's authority by sorting out a fight between two young males. Nikkie was eventually provoked into putting Dandy in his place, but not by simply attacking Dandy directly as any other social animal probably would. First he started an agitated hooting, picking up sticks and stones in a threatening fashion that demonstrated his feelings. Then he went over to the oldest female, Mama, and hugged her in what is a chimp's way of recruiting support. Dandy, clearly aware of what was brewing, ran up and threatened Mama as if to warn her off from backing Nikkie. The whole troop must have been riveted by the social drama being played out between Nikkie and Dandy, because after Nikkie had next gone over to hug Jerome, all the other young males suddenly turned angrily on Dandy, chasing him round and round the pen. Nikkie and Jerome followed the general chase but did not themselves directly slap and harry Dandy as the youngsters were doing.

Such an episode shows how aware all the chimps must have been of the struggle for dominance that was in the air. While the only outward signals were simple hoots, hugs, and threats, a complex web of understanding controlled the behavior of the members of the troop. When matters came to the crunch, it was the whole troop that turned on Dandy, who was given a public humiliation that put a brake on his leadership ambitions—at least for a short while.

Dominance battles catch the human eye because they result in spectacular displays, but chimps show an equally deep level of understanding and communication in more everyday encounters. On one hot day, two youngsters were play-fighting while their mothers dozed in the nearby shade. Suddenly the fight became vicious and the two mothers, while wanting to intervene, were apparently anxious that they might become embroiled in a fight themselves. One mother nudged the dominant female, Mama, in the ribs a few times to wake her up, caught her eye, and waved a hand toward the two rowing kids. Mama got up, took a step forward, and barked angrily—which separated the youngsters—then went straight back to her siesta. For this incident to take place, the mother must have been able to foresee the consequences of stopping the fight herself and must also have known how she could get Mama to sort out the situation. This is obviously a sophisticated level of thought and communication in an animal without formal language.

Observation of the Arnhem Zoo troop has given many other remarkable examples of chimps recruiting support and being able to draw attention to what they want. The chimps make much use of simple gestures, waving their hand in the direction they want another chimp to look or holding out a begging hand for support, then relying on the intelligence of the other animal to sum up the situation and react. Some chimps even develop their own special signals. For instance, Dandy would invite another chimp to groom by holding out one arm with the other in an especially pathetic and pleading fashion rather than using the normal invitation of holding out a limp hand and making soft submissive grunts. Mama had a personal sign for saying "no": She would shake her head in a very humanlike way if, for example, she had held out her hand to invite a male to groom and a female wandered over to sit down. She would shake her head until the female moved out of the way, then start beckoning to the male again. Another example of chimps' awareness of the effects of their signals was the way in which Jerome, a few days after suffering a gashed hand in a fight with Nikkie, limped badly whenever Nikkie was watching but walked without trouble when out of Nikkie's sight. It was concluded that Jerome preferred the quieter life when Nikkie thought he was still injured.

These observations indicate that chimps are the most intelligent communicators in the animal world—even compared to other highly social species such as lions, wolves, and monkeys. This level of communication comes from chimps' having a deep understanding of the social world around them, which means that each chimp must be able mentally to model the impact of its own actions on the group as well as being able to guess the intentions of others.

We have already discussed the brainpower needed consciously to represent the physical world inside our heads. We all carry around an inner understanding of the way water is wet, branches bend, and hard rocks hurt when thrown. With this internal backdrop, we can mentally model the real world and so anticipate the results of our actions, such as foreseeing the effects of throwing a stone at a window or swinging on a weak branch. Highly social animals also need to be able mentally to model the social world of their group, remembering such things as who is dominant, who is bad-tempered, and what actions are likely to follow a particular grunt or screech. Because these sorts of things are less predictable and obvious than the events of the natural world, social animals like chimps and humans need bigger brains to cope with the complexity of their social lives. Making tools would probably have taken up far less of early man's brainpower than learning to get on with his fellow troop mates.

Because the modeling of the social world is second nature to us, we underestimate the difficulty of what we do, forgetting how long it takes children to learn to behave properly in society. However, we have only to break the smallest social rules to realize how aware we are of our social surroundings. For instance, imagine sneezing loudly during a wedding service or turning up wearing jeans. Every minute of a person's life in modern society is governed by thousands of unspoken rules about how an individual should behave. Even a street gang, which sets out to break society's rules, is bound by its own set of rules, so, for example, a gang member can swear at a passerby but risks a cuff if he swears at his gang leader. Similarly, it might be acceptable for him to wear greasy denims but not a silk jacket.

When we consider the huge variety of human cultures, each with its host of social rules to be obeyed, we begin to see the true measure of human intelligence. Very few of these rules are written down, yet even those people who choose to ignore them still know them. Every human head carries a tremendous baggage of knowledge about how people should behave. We may believe that our brains are swollen with facts about the history of the Roman Empire or the geography of Latin America but such schoolbook learning takes up only a few shelves in a mind stuffed with knowledge about the minute details of everyday living.

The only justification for the evolution of the brainpower needed for chimps' social modeling would have been improved breeding success, and we would expect that creating the sort of close-knit social groups seen at Arnhem Zoo allowed chimps to band together in mutual support and raise more offspring to adulthood. Clearly, the very existence of chimps should show that greater intelligence paid off. But as we have seen, before man came along the great apes looked as if they were headed down an evolutionary blind alley. They appeared to be pushing the K strategy of investing more energy in fewer—but better-equipped—offspring close to its natural limits. Before the first *Homo* species came along with food sharing, toolmaking, and the other important advances that allowed hominid mothers to rear many children at once, the great apes seemed to be reaching the likely pinnacle of intelligence.

So far we have been assuming that greater brainpower always leads to better social modeling and a richer, healthier life for a troop of animals, but greater complexity also seems to mean that things can more easily go wrong. So long as evolution is working through genetically programmed behavior, only helpful behavior patterns are likely to be carried on to the next generation. A suicidal animal or one that eats its offspring is not likely to leave many descendants. However, when animals become dependent on a social support structure that they have created themselves, problems can occur.

For example, at the Gombe Stream Reserve, a baby male chimp became unusually attached to his mother. He refused to be weaned away when the mother had another baby when he was about five years old, and watchers noticed that the new baby mysteriously disappeared after three months. The elder offspring eventually regained the full attention of his mother but he never developed properly, and when his mother died of old age a short time later, the now almost adult chimp pined for a few weeks before taking ill and dying.

Even more distressing for the human observers of this wild troop, one female chimp—described as an unusually cold mother—took to killing the babies of others in the group. Over a period of some years she and her grown daughter seized and ate about ten newborn chimps, each time later kissing and grooming the distressed mothers as if the murders were nothing personal. There was no obvious reason for this strange behavior. Perhaps the female's unfortunate childhood—her mother had been injured and unable to look after her normally—was enough to lead to the attacks. Whatever the explanation, the point is that chimps have come to depend on such complicated behaviors to gain a breeding advantage that the dangers of things going haywire start to outweigh the benefits. Just one murderous mother in a troop is enough to wipe out the group in a generation.

Mother fixations, child murders, and cannibalism are not the only way chimp societies can go wrong. The Gombe Stream scientists also saw warfare between two troops of chimps. When a new group moved into the area, its males systematically picked off individuals from the original group, a gang of three to six males beating and bashing a member of the other group for up to twenty minutes. In four years, ten of the old group disappeared, believed to have died from the repeated attacks. This sort of behavior underlines just how closely chimps are related to humans. Intelligence is a two-edged sword: It allows the flowering of a social order that raises the breeding success of a species but complexity also leads to the danger of breakdowns and things going very wrong.

Having looked at the gloomy side of evolutionary progress, we should return to the story of the emergence of language. As we have seen, chimps have a rich social life and good communication. Indeed, with a chimp's begging and pointing gestures, we are getting very close to speech because the intention to communicate a specific idea is there. Early man about two million years ago must have been at least as socially advanced as the modern chimp. After all, *Homo habilis* had a bigger brain and made more use of tools, so he should certainly have been able to point to draw attention to two squabbling children or shake his head to tell another that he was not interested in scratching the other's back.

One of the first steps in teaching human babies to talk is to direct their attention toward things. When a baby is about eight months old, a mother will catch the eye of her baby, who will follow her gaze when she looks across the room to, say, a teddy bear. Once the baby is looking at the bear, the mother can play the naming game—"What's that? That's a teddy!"

We seem to be genetically programmed to be supersensitive to the direction of another person's gaze. Even out of the corner of our eyes, we are aware when someone is staring at us—and so, we assume, thinking about us. We saw with chimps the importance of this sort of awareness in a highly social species of animals, and humans seem to be even more

acutely aware of when they are the object of another's attention—an ability that can be a problem when traveling on a
crowded train with nowhere to look without catching someone's eye.

We are equally sharp at telling when another person has
spotted something interesting and can easily turn to follow
his gaze. If someone we are talking to flicks his eyes across
the room, we automatically glance in that direction ourselves. This sensitivity to another's gaze is an even better
way of capturing and directing attention than waving and
pointing. Such eye-signaling must have been so important to
early man that evolution lent a helping hand, making it easier to tell where a person's eyes are pointed by drawing back
the corners of the eyelids and exposing the whites of the
eyeball.

This suggests that the use of the eyes to establish contact
and direct attention was a major step toward human speech.
Early speech would have been rather short on vocabulary and
grammar, so a glance and a flick of the eyes could be used to
say the equivalent of: "Hey you, look at that." Being able to
focus another's attention would make sure they were looking at the same thing as the speaker. The same image would
fill the other's mind and a few expressive grunts would then
carry a lot of meaning. For example, a nod toward a dying
fire and a disapproving grunt would mean that the fire was
going out and someone had better get some more firewood.

Habilis would probably have been able to say a lot about
everyday life through the intelligent use of chimplike communication, and in the early stages of speech, shared attention and the urge to communicate must have been more
important than the existence of words. But we need to add
considerably more to chimplike communication before arriving at human language. A chimp's gestures and emotional
cries work because both sides can see what is being talked

about. For communication to take place, all the hooting, screeching, and pointing must be about something immediate and obvious to both parties. Real language, on the other hand, is symbolic. Human words are convenient noises that do not mean anything in themselves. Words such as *boat*, *book*, or *hat* do not sound or look anything like the real-life objects they represent. They are just tags and labels.

Once we start using pure symbols to communicate with, we obviously no longer need to have objects in our presence to be able to talk about them. If a chimp wanted to tell you about a boat, it would have to drag you within sight of the vessel and start stabbing its finger at it. A human just has to say the word and the listener's mind provides the rest. As we have seen, each word in a language has been agreed to stand for a net of ideas about something. If we say "book" to another person, we arouse in his mind a net of knowledge clustered around the general concept of a book. The person may even conjure up a conscious image of a brown-covered book with faint gold writing down the spine. If we wanted to awaken a more specific book image in our listener's mind, we might use the word "tome" or "airport paperback," expecting these more precise terms to call forth the image of either a large, dusty old book in a university library or a chunky little book with a shiny cover and racy title.

Each word is like a key that can be reliably expected to unlock the desired image in a listener's mind. As the speaker, we send a string of such keys to trigger a succession of images, quickly sketching a picture in our listener's mind and then reaching inside his head to point to the things we want him to focus on.

A chimpanzee may have a deep understanding of the world and the brainpower to model both physical and social relationships, but that knowledge stays locked away in the gray background of the memory banks until roused by events actually happening in the chimp's presence. Either another chimp draws its attention to the event—like the nervous mother nudging Mama to tell her about the squabbling kids— or a chimp gives vent to its emotions and the others correctly guess the reason for its display—like the dominant chimp hooting at the young male for getting too friendly with the female. But once early man acquired the habit of using symbols instead of waiting for the real thing to come along, he started unlocking all his mental doors. He could not only rouse nets in someone else's mind, he could also trigger nets inside his own head. Suddenly he had access to his memory and imagination. He could stretch backward into his past and forward toward possible futures, so breaking free of the shackles of the present. By learning to speak, man gained the tool with which he would eventually take control of his own mind.

Where did the first symbolic words come from? How did the habit of speaking get started? Probably the first words arose accidentally from the close-knit life that early man was leading. Evolution had pushed early man into tighter social groups so that he could raise more children, a pressure that led to stable partnerships between parents and also to adults banding together in food-gathering parties in order to feed all the hungry young mouths. Chimps, by contrast, do not pair off and they normally split up to forage for their food separately. Troops of early man would thus have spent far more time in each other's company and would have had a lot more opportunity to develop their communication skills. Such close living would have made it far easier for a symbolic use of noises to start.

We have already seen how chimps form their own unique gestures which they use as symbolic signals to other chimps. Mama shook her head as her personal sign for "no"; Dandy held out one arm with the other as his personal invitation to groom. Perhaps chimps also invent their own personal noises, maybe using particular grunts to mean certain things, but such personal noises are not as obvious as gestures to human observers. The point is that it is quite possible for chimps—or early man—to make symbolic use of noises, even if these "protowords" have a fixed meaning only for the individuals uttering them. This use of personal noises would at least be the first step. The next would be for the symbolic noise to be picked up and used by all the members of a troop.

The potato-washing Japanese monkeys and termiting chimps showed how learned behaviors can spread through a troop, but they tend to spread most easily from mother to child. Youngsters are attentive and playful enough to imitate their mother's actions, whereas other adults rarely take the necessary interest. With the close-knit lives led by early man, the children would have had far longer childhoods, during which they depended on the help of adults. Mothers and children would share more time with each other and develop a deeper understanding of what certain grunts or head shakings might mean.

If early *Homo* mothers developed personal quirks in the way they grunted and fussed over their children, these noises could become uniquely full of meaning for a youngster: A certain tone of grunt could be reliably taken as a warning not to stray and another sort of grunt might mean to come and have a bite of a fat grub the mother had just dug up. Then, when the youngster eventually paired off in a "marriage," the close infantile bond would be repeated with another adult. The two adults could bring to the partnership the communication lessons they learned in childhood, teaching each other their unique interpretations of certain grunts. Because both partners would be adults, the use of these primitive words could be refined. A mother-child conversation would tend to be rather one-sided, with the mother making the noises and

the child doing the interpreting. But with two closely attached adults both could start using the same meaningful noise
and once that happened, the word would become increasingly sharply defined in the minds of the two partners.

Imagine a happy *Homo* couple chatting to each other with
expressive grunts and a few symbolic noises meaning perhaps "hungry" or "dangerous." We would not call it speech
but it would certainly qualify as the beginnings of speech.
The children of this couple would quite naturally pick up on
the private language used by their parents and suddenly language would be inherited. Small experiments in the symbolic
use of noises would be passed down from parents to children,
and before too many generations, primitive words would be
spreading right around the troop.

Chimps may be making the first steps toward language before our eyes when we see examples like Mama shaking her
head in a Dutch zoo. Countless generations of chimpanzees
have probably made similar first steps toward speech without their leading to anything, for young chimps do not repeat
the close relationship they have with their mothers when
they grow up and mix with other adult chimps; they do not
pair off with a partner and thus have a chance to develop a
more mature two-way form of conversation. Any private language that emerged would almost inevitably be lost with each
generation, getting trampled underfoot in the rough-and-
tumble world of the adult. But with early man it became
possible for these first steps to be preserved because the stable
adult partnerships needed to raise more children would have
had the fortunate side effect of helping symbolic speech
take root.

Early man did not have to invent the whole of language in
a single generation; it did not take one genius to come up
with the idea of symbolic speech and then teach it to the rest
of the group. Instead, language could evolve gradually over

thousands of generations with each small step forward in the use of symbolic noises becoming fixed, like the useful mutation of a gene. Steadily, over plenty of time, words and grammar would pile up. Language would have started with a simple vocabulary suited only to childhood needs, but once the habit of speaking was established, an adult would soon have invented new words to help in his grown-up life while out hunting, food gathering, or toolmaking. In a beneficial evolutionary spiral, new words would have paved the way for new social behaviors, which in turn would have led to yet more new words. One change would feed rapidly into the other, until one day man would have found himself with full-blown modern speech and a richly developed culture.

We have been vague about what these early words would have been like, but they would certainly not have been like the tidy words of today, with each word carrying a relatively precise meaning. More likely a general grunt would have stood for a very broad idea such as "termiting" or "share the food," serving to focus attention on the general topic of conversation. Extra meaning would then be given through the tone of voice or gestures. For instance, early man might have grunted "termiting" with an urgent questioning tone, rolling his eyes toward a mound he had just discovered. Or perhaps a father would have grunted "share the food" to his son in a firm voice, tilting a chin at a hungry daughter. There would be no need for all the grammar and surrounding glue words of modern speech in order to be understood.

Early man could have got by with this level of conversation for several million years. Indeed, he would have had little choice, since modern speech needs not only a vocabulary but a whole set of physical adaptations that early man had slowly to evolve, the most obvious being the human mouth and throat. The vocal tract of chimps and gorillas is short and poorly controlled, in musical instrument terms; a rasp-

ing bugle compared to the richly varied trumpet of the human voice. Humans can make hundreds of different noises
because their voice box, or Adam's apple, has dropped deep
down the back of the throat and the roof of the mouth has
become arched, giving the human air passage the necessary
shape and length to make a wide range of sounds. Furthermore, humans have developed a mobile tongue, powerful
throat muscles, and strong lips, which allow us to "bite" the
flow of air into the rapid series of vowels and consonants
that make up speech.

Human babies are born with the standard ape vocal plan
and it is only after they are six months old that the voice
box retraces its evolutionary history and descends deep down
into the throat. The reason for this is that the ape's vocal
plan allows apes—and young babies—to breathe and swallow
at the same time because the hole connecting the windpipe
and throat at the back of the mouth can be blocked off so
that the ape can breathe through its nose while drinking or
eating. The descent of the voice box in humans leaves a permanent gap at the back of the throat, which is why adult
humans are liable to choke to death if they talk while eating
and get food stuck in their windpipe. Evolution protects human babies as long as possible by delaying the descent of
their voice boxes so that they can suck milk at their mother's breast without having to break every few seconds for a
breath. But this protection is lost in childhood so that youngsters can get on with the important job of learning to speak.

The physical changes needed for speech show that modern
language must be a very recent invention of our ancestors.
The high arch in the roof of the mouth that helps with voice
production is about the only telltale sign of speech that shows
up on a fossil skeleton. This arch did not start to appear until
Homo erectus arrived about 1.5 million years ago, and even
then the arch was slight. Judging from fossils, modern speech

came along about 100,000 years ago when the earliest examples of *Homo sapiens* were starting to walk the earth.

The development of the human vocal tract was crucial not just because of the range of sounds it allowed man to make but also, and more important, because it made it possible for man to speak quickly. Modern man can speak at the rate of two hundred or more words a minute, to do which, he has to be able to mouth a phenomenal twenty to thirty different sounds a second. By contrast, early man would have been a slow and halting speaker. *Homo erectus* might have been able to spit out only five or six words in five seconds, while modern man can quite easily manage a twenty- to thirty-five-word sentence in that time.

The speed with which we can speak is very important, given the limitations of our short-term memory. We can keep only a certain number of nets brightly alive in our minds at any one time and as each net starts to fade after a matter of seconds, the more words we can cram through this limited working space in a given time, the more complicated the mental picture we can build up. If we were working at early man's speed and could get words through working memory at the rate of only about five every five seconds—five seconds being the average time between breaths—our understanding would be limited. The memory nets triggered by each word would be fading from consciousness too quickly for us to think about something complicated. As if we were using a cup with a hole in the bottom, the meaning of the sentences would drain away from awareness as quickly as we tried to drink it in. We simply could not remember the beginning and end of a big sentence long enough to make sense of it. On the other hand, once we could routinely manage to speak twenty or more words in one comfortable-sized sentence, we had a tremendous increase in the power of language. We could start fitting in all the glue words that help make sentences flow—words like *in, all, the, and, that*—plus the tenses, adjectives, and clauses that fine-tune what we say. Speech could become articulate and grammatical. Man was then able to cram more words into his working memory and literally think bigger thoughts.

The speeding up of speech in modern man was achieved not only by changes to the lips and tongue but also by less obvious, but equally dramatic, changes to the brain. There has been controversy in psychology over whether language is learned or innate—whether speaking is "hard-wired" into the circuitry of a baby's brain at birth or whether children have to be coached how to speak. Both influences, however, are clearly at work. Certainly the genetic or hard-wired changes are important in man. As already mentioned, the reshaping of the vocal tract is a physical adaptation needed for modern language, involving not only structural changes to mouth and throat but also considerable improvements to the areas of the brain responsible for "pulling the strings" of all the muscles used in rapid-fire speech. There must also be some hard-wiring behind the way humans naturally take turns in a conversation, alternating between being listener and speaker. This turn taking to speech is usually taken for granted, largely because it is among the first things we learn in life.

As mentioned earlier, a newborn baby will focus its eyes to look in the same direction as its mother but as well as this gaze-following reflex, babies also show a reflex switching between periods of attention and periods of activity. From as young as six weeks old—long before a baby can make any meaningful noises—it can take turns with its mother in looking and smiling games. This deep-rooted turn-taking behavior means that the baby will later find it natural to listen to its mother's voice and then respond with gurgles of its own. Turn taking paves the way for learning speech.

Another way that we are hard-wired for language is in the

expansion of the areas of the brain used for speaking. Modern man has a brain almost twice the size that of *Homo erectus*, who was presumably still speaking protolanguage, and much of this extra brain probably evolved to provide a foundation for his sentence-forming and speech-understanding abilities. The left-hand side of the human brain is known to be specialized for language, and several patches of brain just behind the left ear have clear roles in speech. These speech areas, which have become visible swellings on the surface of the cortex, fan back from the ear, rather the way the bands of processing in the visual cortex fan forward from a patch at the back of the head. As with vision, the speech areas break up the job of speech into a number of subtasks, but they work together so smoothly that it is as if one structure were responsible for the whole job.

The way tasks are split can sometimes be seen after people suffer stokes, when blood clots damage small patches of the brain. Clots have shown that a coin-sized patch of cortex near the ear is responsible for listening to the sounds of speech and breaking them down into likely words. Another patch just alongside is involved in speech production, turning the glimmerings of a thought into a full-blown and logically structured sentence ready to be spoken out loud. These two brain patches could roughly be summed up as the areas of speech understanding and speech production. There is a connecting pathway between them, and if this is destroyed by a clot, the two patches are left isolated like islands so that victims of this sort of stroke damage can still speak and understand but have difficulty in being able to repeat something they have just heard.

Another problem occurs when damage is done toward the back of the speech-understanding area, near the visual-processing zones. This region of cortex appears to be involved in the skill of reading—where the trigger for understanding comes through the eyes and not the ears. A blood clot here makes it difficult to read or write, though speech can be understood quite normally. This particular problem tells us a lot about the way the brain works. Writing is obviously too recent a skill to have any special help from evolution, yet the brain is plastic enough to form intelligence centers for totally new skills. The ears may have been the original route into the patch of brain that specializes in speech comprehension, but given enough schoolroom coaching during childhood, humans can build a new path from the eyes which plugs into these sound-dominated speech areas. In a wonderfully straightforward manner, the new ability is simply tacked on to the edge nearest the visual regions of the brain.

This shows that the brain would have been flexible enough to cope with the early stages of language without any need for physical improvements, although in time speech proved important enough for evolution to adapt the brain physically to speaking. Given enough time, perhaps, evolution might eventually adapt us to be better readers and we will be able to learn to read as easily and naturally as we learn to speak.

The brain's flexibility is further shown by the way it responds when the whole speech area is wiped out by a stroke. With a lot of training, the right-hand side of the brain can be coaxed into taking over the task of talking, and although the relearned speech is not nearly as fluent as it was with the specialized zones, the brain tissue is still plastic enough to shape itself to new demands.

Returning to the question of whether or not the brain is hard-wired for speech, clearly humans have a number of physical adaptations for language, but in a sense a lack of hard-wiring is also one of our important physical adaptations. As we have seen earlier, the delay in forming the fatty myelin sheaths that fix the brain's wiring in place can be thought of as an evolutionary adaptation. Because myelinization is delayed in the speech centers until a child is about two years old, human children are given several extra years to tune their auditory equipment to the particular rhythms and speech patterns of the cultures they are born into. The combined effect of these evolutionary adaptations is that children are born ready to learn language. They have reflexes that cause them to try to take part in conversations long before they know anything about speech. They have the brain centers and vocal equipment to make rapid speech easy. And they have several years after birth when their brain centers can be molded to society's ways of talking. Given willing language teachers, such as parents and relatives, it is not surprising that babies pick up language so quickly.

The sorts of physical adaptations mentioned so far are easy to pin down, but it is more difficult to answer the question of whether grammar is also something that is hard-wired into our brains. Many people have argued that children have an innate sense of grammar and so will automatically construct orderly sentences. However, it now seems more likely that all that is wired into humans is the ability swiftly to pick up the grammar of the cultures they are born into. Also, to a large extent, the basics of grammar follow naturally from the human-centered view we take of the world around us.

Whether innate or learned, the workings of grammar are very important to our story of the mind. Grammar lays down the rules by which we assemble the chains of words that parade through our conscious plane, so that in many ways our grammar and our style of thinking are very much the same. Our higher mental abilities have been likened to towers built out of language and erected on the foundations of the natural animal mind, so if words are imagined as the building blocks for these edifices, grammar is the style of architecture. Some types of grammar may give us crude and stumpy towers, others ornate and flowery buildings, or tall and lean skyscrapers. Given that grammar has such an impact on our style of thinking, we should take a closer look at it to see exactly what shapes it.

Language teachers often talk about grammar as if it were a biblical set of rules carved on tablets of stone, but rather than a list of commandments, grammar is fluid, and changes as fast as cultures change. The common feeling that grammar is or should be fixed probably comes from the mass teaching of writing during the last couple of centuries. Educators have tried hard to impose a standard style of spelling, punctuation, and sentence structure on the English-speaking world, an attempt at uniformity that makes considerable sense. With spoken language, listeners are always present and we can repeat or expand upon what we are saying if it is unclear; indeed, the sloppiness of most spoken speech is obvious if what is said is literally transcribed onto paper. The written word, getting no help from the speaker, needs to be precise and orderly to be sure of being understood and this has led to an exaggerated sense of the strictness of grammatical rules.

The halting protospeech of *Homo erectus* probably had little recognizable grammar. *Erectus* would likely have grunted a general word like *termiting*, then relied on gestures and the context to fill in the gaps in understanding. Conversations in modern English, however, are made up of whole sentences—their length limited by the number of words that can be said comfortably in one breath—and each sentence has a complete logical structure. It is a tale with a beginning and an end, always having a subject, a verb, and an object, that is, a doer, what the doer is doing, and what he is doing it to. These three basic components of a sentence are usually escorted by a flotilla of modifying words that give extra shade and emphasis to what is being said: adjectives, adverbs, tenses, and qualifying clauses gather in clumps around the subject, verb, and object, to give them a more precise meaning. Holding these three basic clumps of words together are the glue words—such as *if, but, however, and, that, then*—which are also often used to link several sentences when we are trying to build up a larger picture in conversation or writing.

There are obviously a variety of sources for this structure. Some of the orderliness is a reflection of the patterns of cause and effect we see in the world around us. And some parts of grammar are an arbitrary point of style adopted by a culture. But the true root of grammar is probably a lucky fluke stemming from the way language has to spill out of our mouths in a steady two-dimensional stream of noises.

Once early man developed the first symbolic words, he had no choice but to speak them one at a time. While he could quite easily make several gestures at the same moment—for example, rubbing his stomach, rolling his eyes, and pointing at a tasty fruit hanging from a branch, all at once—if he wanted to express himself in words, he had to string the different words together in an orderly chain. The serial nature of speech would probably, in fact, have been a serious drawback to early attempts at language. Man's ancestors would likely have tried to get around the handicap as much as possible by resorting to the three-dimensional world of gestures—pointing and face-pulling to give extra clues to what they were trying to say. But eventually the two-dimensional limitations of speech turned out to be a great advantage. Man was forced to put his words into some sort of sensible order and the beginnings of grammar would have been born. This meant that he no longer needed gestures to expand the meaning of what he was saying, and it is noticeable that although we still use a lot of gestures or body language when we speak, we tend to disregard them and concentrate on what is actually being said. It is as if evolution wired up *Homo erectus* to make gestures but modern speakers no longer need them now that we have proper speech.

Once man started stringing words together in orderly sentences, he could also start creating modern words. As has been seen, protowords probably described very broad ideas such as termiting, the precise meaning coming through gestures and the context in which the word was used. When man got into the habit of using a string of words to get across what he had to say, these general words could become more narrow in their meaning. The general grunt of "termiting" could be stripped down into its components so that there would be different strings of noises for saying, for example, "I have been to dig for termites" or "About time you went termite digging, son." Man could replace gestures and context with a host of extra adjectives, tenses, and glue words. The serial nature of speech would have forced man to adopt some sort of sensible order for his words. Not surprisingly, the order he used was a reflection of the natural order that man could see in the world.

The way that man breaks his sentences down into the basic units of subject, verb, and object fits the active way that man views life. When we look at the world around us, we are always looking to see who is doing what to whom. The whole assumption of human grammar is that there is an active person at the center of the scheme of things who goes out and does something to the objects in the world around him. Such a human-centered grammar, where everything has a cause and an effect, has a logic perfectly designed to express everyday ideas, but it is not always the most accurate way of seeing life. Such logic is too simple to capture properly the complex feedback relationship of something like evolution, where an effect can also be its own cause and where there is no grand designer causing things to happen but where stable patterns emerge as what works outlasts what does not. However, despite the difficulties of our human-centered grammar, its simple "doer and done to" logic still covers at least 90 percent of what is important in our lives.

The way that grammar probably developed in human speech is, then, that sentences had a natural two-dimensional order and a length determined by the limits of working memory; a natural subject-verb-object structure then developed that mirrored a human-centered view of life. After that, the rest of the rules of grammar could be left to the local conventions of different cultures. There would be no real reason for a particular grammatical rule other than it had become customary. This is probably why the modern world has hundreds of different tongues which all handle the minor details of speech quite differently. It does not much matter whether adjectives come before or after a noun or how many different tenses exist for verbs—so long as everyone in a culture is following the same rules.

The Hopi Indians of North America, for instance, do not have the vast array of tenses of a language like English, which allows English speakers to place an event precisely in time. English speakers are very sensitive to such distinctions as whether an event is going to happen, has already happened or what it would be like if it were to happen. The Hopi language, by contrast, has a sense of timelessness about it. Hopi Indians are more interested in how sure the speaker is of his facts, and verbs vary according to whether the speaker is talking about a current event, a remembered event, an expected event, or a broad generalization about such events.

The Navaho Indians also have a grammar that sounds odd to the English-speaker's ear. They link their nouns and verbs so that the doer is seen not so much as causing events to happen but rather as becoming involved in various states of being. For example, "an Indian skinning a deer" is expressed in Navaho as "an Indian taking part in the act of deer skinning." Rather than being the bold originators of an action, Navaho Indians talk as if they have become involved in a class of activity that has gone on since the beginning of time. Another idiosyncrasy of Navaho speech is that verbs are varied to match the shape of the object involved, according to whether they are long, tubular, or sheetlike. For example, a sentence might be rather like saying in English: "I will (tube)burn my spear and (sheet)burn my tent." These forms of grammar still have a human-centered order of subject, verb, and object—if perhaps a little less human-centered than the European norm—but they show how the set of rules governing English grammar is just one of many different ways we could have structured our speech. The basic structure of grammar reflects the human viewpoint but the detailed rules are an accident of history.

To summarize the evolutionary history behind language, we see that our *Homo* ancestors probably started out at the same level as the chimpanzees, where communication relied on members of a group being intelligent interpreters of each other's cries and actions. Then early man developed symbolic noises that were trapped and inherited through the close-knit relations he had as an adult. These first words might have had a blurred general meaning but they proved useful enough for evolution to start improving our ancestors' vocal equipment. Speech continued to improve steadily until it was fast and rich enough to spark a true explosion in language.

We have no real evidence for exactly when either the first stage of protolanguage or the second stage of full-blown language occurred in the course of man's history, but it seems certain that by the time *Homo erectus* came along about 1.5 million years ago, he would have needed a reasonable language ability to go with his making of tools, use of campsites, and control of fire. However, *erectus* developed only slowly: The heavy stone axes he carved stayed unchanged for nearly a million years, which suggests that his whole lifestyle stayed much the same and that a lack of specialized vocal equipment might have held him back.

By about 200,000 years ago, Neanderthal man had arrived. Rather a mystery, he had a brain as big as modern man's—in some cases even bigger—and possibly showed the first signs of culture and superstitious beliefs. There is some evidence that Neanderthal man buried his dead in graves full of flowers and had tribal ceremonies involving the skulls of the giant ice-age cave bear. If this level of symbolic thinking occurred, it would imply that Neanderthal man had a reasonably advanced form of symbolic communication—that is, speech— to underpin the ideas. However, Neanderthals still had a weak apelike chin and relatively low arch to their palates, which could show that they had not yet developed the vocal organs needed for clear and rapid speech.

The evidence to date suggests that the Neanderthals were probably a sturdily built northern offshoot of the hominid family, one specially adapted to a hard hunter's life on the edge of the glacier fields. At the same time that the Neanderthals were dominant in Europe, a more lightly built version of early man—*Homo sapiens*—was evolving in Africa or the Middle East. This skinnier branch was the one that eventually developed modern language and gained all the mental advantages that went with rapid speech. This development may have gone on for thirty thousand years or so, leaving little trace in the fossil record, but by about forty thousand years ago the fossil record shows clearly that modern man had arrived.

Tribes of people identical to modern man started to spread rapidly across the world, replacing Neanderthal man in the process. The quick migration of modern man shows that he was even more adventurous and adaptable than either *Homo erectus* or Neanderthal man and he not only quickly colonized Europe and Asia but traveled for the first time to the Americas and even reached Australia, a journey needing a boat trip.

The clear sign that this tribe of men who emerged forty thousand years ago had a modern self-aware mind is the explosion of culture that took place as soon as he made his appearance. *Homo sapiens* started to make delicate flint arrowheads and carved bone fishhooks. Far more telling, he had religious rituals and made the first use of art and decoration. The murals of animals in French caves are well-known examples of prehistoric art, dating to about twenty thousand years ago, but as well as these paintings, many other types of art started appearing, such as bead necklaces, clay statuettes, and chiseled drawings on tools. Clearly, man must have been self-conscious and articulate by this time. He had made the break with the present tense and lived in a mental world of his own creation.

When modern man emerged, the last ice age still had a frosty grip on the planet, but about ten thousand years ago the hard life of the ice ages finally started to ease. The rising temperatures not only pushed back the glaciers to allow forests once again to cover the temperate zones but also brought back the rains to make the tropics lush and fertile. Summer had finally arrived and *Homo sapiens* was ready to take quick advantage of the vast improvement in the world's climate.

When the last ice age ended, there were probably only about ten million *Homo sapiens* in the world and they all lived a hunter-gatherer life-style. Such a way of living meant that tribes had to spread out around the landscape if everyone was to get enough food. This isolated life-style would obviously have limited the language of these early men, who by this stage would have had a robust grammar and rapid speech, although their simple needs would not have demanded too much complexity or abstraction. Also, with so many small wandering bands, there were probably hundreds of dialects of a similar tongue. There would not have been any large concentrations of people to push the development of a language along.

However, one of the features of symbolic speech is the ease with which it can be changed and improved. English speakers, for example, use only about forty-five different sounds to make up words—the different syllable noises like the *ch* in *church* or the *ng* of *sing*—yet these forty-five basic sounds have been combined in enough different ways over the past few thousand years to fill a dictionary with half a million words. Also, because words are symbols, there is no limit to what they can be tagged on to, so that we can name things we cannot see, like an atom, or put a one-word label on an impossibly complex idea such as the universe. Humble beginnings would not have slowed the progress of language once the ice ages had lifted and man started to flourish.

The easing of the weather made it possible for man to give up the hunter-gatherer life-style he had followed for several million years and turn to farming. The first move was the domestication of animals, which probably had its beginnings about fifteen thousand years ago, during the ice ages. Some tribes lived by trailing around after herds of reindeer, getting hides for clothing, horn for tools, and almost all their food from them. The change in climate a few thousand years later opened up pasturelands across the Middle East, and about nine thousand years ago man learned to capture and tame his own herds of sheep and goats. This led to an even higher standard of living and a more ordered culture, but even though man no longer had to go out and hunt, he was still leading a nomadic life, which would have limited the advance of culture.

A more settled existence came with the cultivation of grain crops. Grasses like wild wheat and barley grew in lush pastures with the change in climate at the end of the ice age and at first man would have harvested this natural bounty. A family could gather enough wild wheat in three weeks to last them the year—once they knew the secret of how to store it sealed in the ground in grain pits. The primitive farmer probably then noticed how grain spilled outside his hut soon sprouted. From there, it would have been a short step to growing crops himself.

Once man started growing his own food, many changes would have automatically followed. A farming culture can support a hundred times more people on the same land than the hunter-gatherer life-style. When the ice age ended, almost overnight people started putting down roots, gathering possessions, and building cities. This in turn led to social hierarchies and specialist jobs. The settled groups also had to learn to defend themselves against outside raiders and develop laws to control the citizens within the city walls.

One of the oldest known towns is Jericho. This was an oasis visited by nomadic herdsmen eight thousand years ago, but by six thousand years ago it was a town protected by fabled walls, covering ten acres and with two thousand inhabitants. Similar towns and then cities sprang up all over the Middle East. By about four thousand years ago the great river-valley civilizations had started, with the fertile river plains of the Nile in North Africa, the Euphrates and Tigris in the Middle East, the Indus in India, and the Yellow River in China all supporting large city-states. Right around the world, there was an explosion in human culture—and in the language and thought that goes with it.

The climatic changes at the end of the ice ages thus created an opportunity that man, now equipped with language, was ready to exploit. Language led to civilization and the new civilizations in turn fueled further progress in the speech and mental processes of humans. The rise of privileged classes within early civilizations—such as priests, nobles, and craftsmen—created groups of people who had the time to develop new spiritual values and fresh ways of thinking. Freed from the daily grind of food gathering, these leisured classes developed the writing and counting systems that were to have a great impact on the mind. But before looking at the detailed effect of the changes wrought by civilization, we shall return to the very first effects that language had on the mind of early man and look at what must have been going on inside his head once he developed his new habit of speaking.

SIX

Strange Voices in the Head

About 100,000 years ago, Neanderthal man could probably speak. He may have had only a limited vocabulary and a rudimentary grammar but he could probably string together enough words to tell someone to shift over at the fireside or shout that a saber-toothed tiger was prowling in the far bushes. However, while Neanderthal man might have been able to talk to others, could he have talked to himself? Had language become advanced enough for him to internalize speech and have a little voice chattering away inside his head?

To modern humans, their inner voice is a deep-rooted part of the mind, chattering on about anything that plays across their awareness like a running commentary on a televised sports match. It bubbles up within, commenting on the sights we see and memories we recall, monitoring all that passes through the mind and clothing it in words. For example, if we look around the room in which we are sitting, it may suddenly come out with a general comment such as "This place is messy; there are books and papers scattered everywhere." Or if we peer closely at something, it will chime in with the name of the object.

This inner voice is the constant companion of our thoughts and sometimes it feels like the very core of our being. But what is it really? To answer this question, we need first to recap what has been said about the general mental landscape against which the inner voice plays.

Our conscious plane is defined by all the fleeting nets of nervous activity racing across the brain surface at a particular moment. This conscious plane is not uniformly bright but has sharp impressions of the things in central focus and a general background glow of awareness for things at the fringes. The feeling of being aware comes from a low-level buzz of recognition—like a quiet sigh of "aha"—for all the sights, sounds, and body sensations that surround us. If we are sitting in an armchair in our lounge at home, we have a net of knowledge about the room—what it looks like and what sort of things are likely to happen in it—and as long as nothing extraordinary happens, like a car backfiring in the street or a rabbit suddenly skipping across the floor, our awareness of the surroundings fades away to a gentle buzz of recognition. We sit there with a reassuring feeling of continuity, knowing we are living in one moment that stretches back into memory and forward into future possibilities.

This general orientation to our world serves as the backdrop to the sharper impressions that form the central focus of our attention, that is, the immediate sights and sounds of whatever it is we are concentrating on. If we follow events in this bright central awareness closely, we can see how it is filled with momentary distractions. A sudden itch on the leg, the movement of a crawling fly, or the noise of a passing car might catch our attention briefly but when they have been

rapidly assessed as unthreatening or uninteresting by the usual
mechanisms of recognition, we return to concentrating on
the task at hand.

Chimps and other higher animals must have this same raw
consciousness with its sharp focus at the center and general
glow of awareness at the fringes. The difference from human
consciousness would be that all the nets making up an animal's plane of consciousness are tied to the world of the moment. Humans, on the other hand, can also fill their
consciousness with memories and imagination, and they
manage this trick by using the one artificial component of
their minds—the inner voice.

This voice inside our heads seems to well up like a spring
of water that cannot be capped. It bubbles on and on in an
endless stream of thoughts and suggestions, and no matter
how hard we try, we cannot shake it off. If you attempt right
now to empty your head of all thoughts and sit quietly with
a blank mind, you will probably last barely a second before
catching the faint scratchings of your inner voice in some far
corner of your mind. And as soon as you notice it, the voice
will come bounding back into the limelight like a grateful
dog returning to its owner's side. Even if you try desperate
measures to block out the voice, such as repeating a monotonous phrase or humming in your head, it will still creep
back in and you will suddenly catch yourself saying something like "This is bloody stupid" or "Is it still there?" Once
the faintest trace has caught your attention, the inner voice
will click back into focus and start spilling out its endless
stream of words again.

There is nothing in fact particularly mysterious about this
inner voice that we can never switch off. It is simply the
region of the brain that has been wired up to produce speech
getting on with its job. This speech center—the coin-sized
patch on the left hemisphere—will go on churning out phrases

and fragments of thought regardless of whether we are actively speaking or sitting quietly with our mind idling in
neutral. If we try to ignore the voice by mentally looking
elsewhere, it will nag at our attention just as much as will a
fluttering rag seen out of the corner of our eye. If we try to
drown it out by humming or mechanically repeating a meaningless word, we can go on for only so long before realizing
that the inner voice is still there trying to form sentences. It
cannot be switched off, just as we cannot switch off our vision or sense of smell. We have a conscious control over the
muscles of our bodies, and so can paralyze the muscles of our
throat to prevent our talking out loud, but the brain's speech
center cannot be shut off any more than the brain's visual
areas can be stopped from seeing when our eyes are open.

The way that the speech center puts together its sentences
is a more complicated issue. It is clearly not done consciously since we become aware of the words and phrases
only as they emerge. When we speak we are sometimes
amazed at our own wit or eloquence and wonder where the
words have sprung from even as they pass our lips. Before we
have said something, we really have no inkling as to how it
is going to come out. We can rehearse a sentence in our minds
before repeating the words out loud, or we can weigh our
words carefully and stand by ready to step in to replace an
unsuitable word in the split second before our mouths manage to spit it out, but the speech center itself seems to be a
black box in our heads which it is impossible to enter.

If we analyze the process of forming a sentence, the
impression is that we start by filling our minds with vague
feelings of what we want to say. We begin with an image or
germ of an idea in our minds and this raw material gets fed
into the hopper of the speech center, after which we effortlessly crank out a string of beautifully chosen words. For example, if we describe the way we made ourselves a cup of

coffee this morning, we would first stir up a mental picture of the experience, perhaps seeing ourselves holding a steaming mug, and from that image the speech center would start scooping up enough thought to fill a descriptive sentence. Perhaps the first point we feel we should mention is the time of day when it happened and we find the sentence bubbling up that we drank the coffee in a rush because we were running late for work.

Looking at this first step in detail, even to get this far there had to be quite a lot of back and forth between the speech center and our attempts to recall the episode. The speech center would come out with half-formed questions to prompt us to dig out the important details, the inner voice perhaps running something like "Well . . . (what did we do first?) . . . time . . . (we could start with that) . . . late? . . . (yes, we were in a hurry)." After this to-ing and fro-ing—taking perhaps all of a second—we are finally ready with a full-blown sentence. Our speech center grabs the strings to the muscles in our throat and we find ourselves telling the other person about gulping down a cup of coffee as we rushed out of the door.

Usually when the words finally pop out of our mouths, they serve to make our originally vague feelings seem precise and clear, giving a sharp outline to our initially woolly thoughts. The newly formed sentence then sends our mind racing down fresh avenues of thought, calling up new images and sowing the seeds of the following sentence. Speaking is thus a rather complicated process. The speech center seems to prod our imagination and memory with the half-formed questions of the inner voice. This rouses the woolly mental images that the speech center then turns into an appropriate chain of words. The speech center seems to hang over our raw minds as if it were a bullying teacher trying to extract an answer from a stupid pupil, pushing and probing until the

inarticulate student comes up with a few rough-cast memories. The impatient teacher then seizes the mental image and
turns it into an elegantly polished sentence, which is repeated for all the class to hear.

How can the speech center show such intelligence? Is it
really the hidden core of our being that has so far eluded our
searching? The answer is that the speech center is only one
of many intelligence centers in the brain. It has no supernatural powers of self-awareness or reasoning. Like the patch of
nerves that moves our hands or decides how far away a cow
is standing, it is just a mesh of nerves trained to do a particular job. To see how our speech centers become trained in
the habit of clothing images in words, we can look at how
every child is taught to speak.

Children, as has been seen, are born ready to talk, with
the raw brain power to learn, the vocal equipment to experiment, and several years of plasticity to get used to the
rhythms of speech. Parents are natural language teachers and
from a child's earliest days, it is surrounded by adults trying
hard to communicate with it. These adults are prepared to
go to any lengths to get through and will allow themselves
to look ridiculous by talking baby talk.

Baby talk may sound silly but its exaggerated tones and
repetitive questioning are exactly what is needed in the early
stages of learning speech. The typical first exchanges start
with the mother catching her child's attention and saying
something like "Who's a good little boy, eh? . . . Who's a
good boy? . . . [tickles baby's stomach and baby goes "gah"]
. . . yes, you . . . you're a good little boy." The mother carries the conversation along on her own but treats any noise
by the infant as a valid attempt to talk. She naturally focuses
on the simplest and most obvious topic—the baby itself—
and hammers away until the words and speech rhythms
become familiar. As the training progresses, the mother be-

comes more demanding of the responses she expects and also turns the focus onto the outside world. This leads to the familiar naming game which gets children accustomed to the idea of pinning a label on everything they see, so that having got her child used to responding in a conversational-type manner, the mother teaches it to say, for example, "doggie" whenever a dog wanders into view.

It should now be clear where the habit of clothing thoughts in speech comes from. The speech center is first taught the relatively simple task of forging a mental link between real-life objects and certain vocal sounds that adults call words. Every time a child sees a dog, the image funnels in from its eyes and is splashed across its conscious plane. At the same moment the child has its mother saying "doggie" in its ear and the sound creates another conscious net. We have seen how it is natural for two firing nets to form a link, so after mother has said "doggie" a few times, the sight of a dog should by itself be enough to prompt the child to become conscious again of the word "doggie." Before long, with encouragement from mother, the child will also squeal "doggie" every time it spots a dog.

It usually takes some time before a child gets the use of the word exactly right. To start with, it may call all four-legged animals—or even daddy—a dog because it is still so young that it has a rather unfocused net of knowledge about animals. The label is attached to anything hairy that moves. Later, the child learns that only dogs should be called "doggie." Words sharpen up the child's perception of the things around it and puts its knowledge of the world into neatly defined nets.

By the time children are two years old, most of them are skilled at using names. Adults may have to teach children to make the first associations between visual sensations and the appropriate words symbols, but children quickly catch on to the trick and learn new words under their own steam. They will also move on to naming more abstract words and ideas, learning to describe the position of objects through using words such as *down*, *on*, or *in*, and learning to comment on life by using words such as *bad*, *no*, or *oh*, *dear*. Once having got into the improbable-seeming habit of linking sights with arbitrary word sounds, a child finds it easy to keep extending its vocabulary.

The next step is to learn to string words together. Instead of crying and holding its arms out to be picked up, a child starts putting together a protosentence like "Mummy, up!" By age three and four, children will speak complete sentences formed according to the conventions of grammar. Grammar is picked up by children in much the same way that they build up a general net of knowledge about anything else in the world. They listen to many examples of adult speech and then form a blurred mass of experiences from which the broad ideas stand out. After hearing a lot of talk, children get a general feel for how to fit tenses to verbs and what sort of word order sounds right to their ears. They do not have to be taught the rules of grammar. Rather, the rules become obvious after they have soaked up one or two years' worth of memories of other people's conversations. The general rhythms of a language eventually etch the processing surface of the speech center and without any deliberate effort, the individual grows up trained to assemble chains of words in a standard way—and to talk and think in the standard rhythms of the culture in which he is brought up.

This process of learning grammar can be seen from the particular types of mistakes four-year-old children make in their speech. Instead of being sloppy about grammar, like an unsure toddler, four-year-olds become too rigidly correct in applying the newly learned rules, saying things like "I seed him" instead of "I saw him" or "He runned there" instead of "He ran there." They try to use what should logically be the past tense rather than the illogical form actually required by English. This happens even if the child had been saying "saw" or "ran" a few months earlier when it was not so conscious of the demands of grammar. Clearly the child's speech center can become so tuned to the broad rhythms of language that it tries to stamp every sentence with the same mold. The grammar has become part of the way the child thinks.

The speech center created in every child is stocked with a vocabulary and a feel for the grammar needed to form sentences. It is then trained to react to the parade of images passing through consciousness by cloaking them in the appropriate words. But how do we go from learning to speak to the people around us to learning to speak to ourselves in the privacy of our own heads?

With the toddler learning to say "doggie," both the dog and the word were at first objects in the outside world; the child saw the dog and then heard its mother making a loud noise. Later, the child sees the dog and hears itself mouthing the same noise—getting encouragement from mother when the noise sounds like "doggie." By normal associative learning, the child's brain has formed a link between the sight and the word, just as if it were a rat being trained to push a lever. Later still, the child takes the important step of seeing the dog and silently saying the word in the privacy of its own mind. With adult encouragement, the child learns to internalize speech by freezing the muscles of its vocal cords and stopping words from escaping its lips. The brain center that handles the word pronunciation by tugging at the strings of the voice box will still buzz with all the correct messages, but as when we imagine lifting a hand yet stop ourselves from doing so, the child will have learned to imagine saying the words without actually making any noise. The child will have finally created its own inner voice.

If we listen to children at play we can see the process at work. More than half of what young children say when engrossed in play could be described as thinking aloud. Rather than trying to communicate with the other children around them, they are putting together a thread of thought in their own minds. For example, a child building a tower of blocks might say, "Where's another long one? Here it is. It doesn't fit. Oh, dear." This sort of prompting of the mind should be familiar from our adult attempts to gain control over memory and imagination. The child is learning aloud the habits of thought that it will later learn to use in private. This thinking aloud goes on from the age of about four right up to eight or nine, by which time children have usually learned to speak silently in their heads.

Coming back to the question of whether Neanderthal man of 100,000 years ago had the inner voice so necessary for a modern self-conscious mind or could speak only aloud to other Neanderthals: It seems unlikely that there was any physical reason stopping him from forming an inner voice. But it is also quite likely that his culture was not advanced enough to encourage it to happen. Modern humans are actively encouraged to think silently. People who think aloud in public are treated as mad, for spoken words are acceptable only when they appear to be aimed at another person. There is clearly enormous social pressure put on us to learn the trick of silent speech and moreover much of modern life depends on having an inner voice. All our mental extensions, such as personal memory and imagination, are based on the control that comes with an inner voice and, as we shall see, so too are other human abilities such as higher emotion and self-awareness. All these pressures combine to make it essential for modern humans to learn to internalize speech.

Neanderthal man, however, may not have faced such pressures. Perhaps he used words far less often. Like a toddler of today, he may have needed words only occasionally, when faced with a problem or when he needed to get another Neanderthal to do something. He could have got by with childlike thinking aloud, never needing to learn to speak inside his head. After all, the original purpose of symbolic language was only to let early man communicate with others in his tribe. It was luck that speech turned out to be a method of communication that man could turn around and use to communicate with himself.

This raises the point of just how lucky humans were to strike upon the spoken word as their method of communication. We have looked at how some advanced animals, like chimpanzees, can use emotional cries and symbolic gestures to communicate with others. However, while cries and gestures do the job quite well, neither could easily form the sort of inner structure that language has become.

Emotional cries are not learned and so animals cannot properly control the way they come out. Evolution has simply made sure that a social animal like a chimp makes its inner feelings plain to other chimps through various screeches and whimpers. Cries, too, are not symbolic. They can be interpreted only from the context in which they are made. Given these crippling limitations, it is hard to imagine holding a meaningful inner dialogue with ourselves through a series of imagined shrieks and grunts.

Gestures, while they can be deliberate and symbolic, like Mama shaking her head, cannot be monitored in the same way as words. We cannot see ourselves as others see us; we have no clear picture of exactly what we are doing when we pull faces or flap our hands about. Also, we can make a number of gestures at once, so that they do not have the same two-dimensional limitations that proved so fruitful for getting language off the ground. Some humans have, of course, learned to use a gesture language. The deaf can talk fast and fluently through a language of hand signals, facial expres-

sions, and lip reading, and some people who have been totally deaf from birth—which is rare—say that they tend to
think using a series of gestures as their equivalent of an inner voice. Or, if they are good readers, they can use an imaginary flow of printed words. However, their sign language is
a translation of spoken language and it seems unlikely that
a speechlike sign language would have evolved by itself.
Speech had key advantages in that it was not only under conscious muscular control but we could also hear our signals
as clearly as our audience.

When we consider that the grammatical order of language
was another accident—an order forced on man because he
could speak only one word at a time—the emergence of man's
mind can be seen in its true perspective. After several billion
years of slowly evolving life on earth, a small experiment in
communication by a fairly insignificant two-legged ape suddenly led to the surprise eruption of self-aware man.

To sum up the story so far: Biological evolution got man
as far as his natural mind. Like a high-horsepower version of
an animal mind, the three-and-a-quarter-pound human brain
had a powerful ability to perceive, recognize, learn, and be
consciously aware. Then along came the small experiment
of language with its surprise features. Speech was symbolic
and not tied to referring to things existing in the present;
thus it freed humans to talk about the past and future. It was
compact, so words could be used as place markers to shuffle
huge chunks of knowledge around in the limited working
space of conscious awareness. Speech could also be internalized and used as the inner organizer of what went on in the
brain, replacing the outer world as the sole driver of the mind.
Language, and the inner voice that eventually went with it,
proved to be the building material for a whole range of extensions to the brain's natural facets.

The new artificial abilities of the mind included those already mentioned, such as personal memory, factual memory, imagination, and word-driven chains of thought, but there are many others, like conscience, word-driven attention, higher emotions, personality, and self-awareness. Before looking at these, we should first see how such artificial facets came to be created. They obviously needed some shaping force and did not just emerge fully formed as further chance side effects of language.

Probably, however, not much more needed to be done. After the billions of years it took for evolution to go from the simple reflex-driven nervous system of a worm to the complex conscious brain of the ape, the artificial parts of the human mind were added in a comparative eye blink of time. This indicates that they must have been simple to add. The natural foundations of the human mind are rather like an iceberg with huge unseen depths. Every second, our brains are dealing with a tidal wave of information flooding in from our senses, muscles, and organs. Our eyes alone each have 125 million light-detector cells firing off messages that demand our attention. Somehow, with every nerve clamoring to get through at the same instant, this torrent is filtered and condensed as it races toward the brain. The images and feelings that are finally splashed across the conscious plane are the distilled essence of the moment.

On top of this pyramid of processing, language and the artificial facets of the human mind lie like a thin crust, but this thin crust has tremendous power because it can turn the processing pyramid on its head. Animals are stuck in the present, with each new wave of sensation funneling its way up to the summit and washing away earlier waves, but language sits atop the pyramid and can drive thoughts fanning back down again. A couple of words, like "blue crocodile," can spark an image in our mind; they awaken nets in our memory banks and send images fanning back across our visual area. In effect, language has turned the funnel into a trumpet. The thin crust meant that the brain was no longer driven one way by the outside world but started responding to the word-driven chains of thought taking place inside.

Nevertheless, while it may have taken only the small experiment of language to stand the brain on its head, these new habits of thought would still have needed a shaping hand. There must have been some reason for us to develop such habits as reviving personal memories or stepping back self-consciously to observe our own minds at work. Clearly there was a new form of evolution creating these extra mental abilities—cultural evolution.

Cultural evolution is a simple idea and works just like biological evolution, except that instead of the inheritance of physical changes, it involves the inheritance of useful behaviors and patterns of thought. The spread of potato-washing in Japanese monkeys and termiting in chimps were crude examples of cultural evolution in animals. Another example is the way some baboon troops in Africa have discovered that they can safely eat scorpions after batting them around on the ground until they are stunned. Neighboring baboon troops who have not learned the trick can only teach their youngsters to flee in fright from potentially tasty scorpion meals.

These examples of cultural inheritance are impressive but they lack an essential ingredient of true evolution: There is no equivalent of the genetic material that underpins biological evolution. There are no genes to code for a lot of very slight changes in behavior so that monkeys could evolve toward potato-washing or scorpion-rolling through a series of small intermediate steps. One monkey genius might get the trick right in a single go and then pass it on to others, but a thousand monkey generations could not gradually build their way up to a complex feat of thought because there would be nothing to fix all the in-between stages. However, when language came along, it provided early man with the equivalent of genes for cultural evolution.

Because of the perception-sharpening power of words, speech can capture the most delicate and precise shades of meaning As we have seen, naming something makes it stand out mo⟩ clearly from the surrounding background. If, for example, ea⸱ man got into the habit of calling his resting spot for the n⸱⸜ a "camp," this would make its imaginary boundaries stand out more distinctly in his mind. Suddenly it would not be just a forest clearing with a log fire and thirty slumbering bodies but a home with the ground kept free of rubbish and children stopped from straying beyond an imaginary perimeter. The word *camp* would not only make that day's resting spot stand out in each individual's mind, it would also be a focus for all sorts of memories and associations. Once a campsite became a defined thing in early man's mind, he would have been able to remember places he had stopped at before. Also he could create a mental checklist of rules for setting up good campsites and remember the various rituals needed to make a camp safe from evil spirits.

Having a word gives sharpness to man's impressions and ties them to bundles of knowledge stored in the memory banks. With each passing generation, the words and their baggage of ideas would be refined and polished with use. A camp may have meant a convenient forest clearing to start with but later may have come to mean a place that needed to be blessed with ceremonies, where the oldest person got the hut nearest the fire, or the place where a tribe would return each spring to hunt game.

Eventually, a troop of early humans would have evolved many different words with rich webs of meaning behind them. A complexity of thought that would have been beyond any individual would be evolved steadily over many generations. What would be passed on would not be just the meaning of words but also ways of living and behaving that had been remembered through the use of words. Gradually every tribe would form its own rich heritage of ideas, attitudes, and beliefs. As each new generation arrived, the impressionable young minds would first be taught to speak and then taught all this stored up custom and lore. Any small extra advances in culture and understanding made by the new generation would in turn be added to the baggage of ideas passed to the next generation down the line. With each small step fixed by the gradual expansion of individual words, simple ideas about how to do things or what was morally right could have steadily evolved into complex ideas.

As with biological evolution, cultural evolution would have acted with blind statistical force. What worked would tend to outlive and outnumber what did not, so that the ideas and behaviors that proved useful to man's survival would be the ones most likely to be passed on to future generations. It is easy, then, to see how useful behaviors could have evolved through cultural evolution but perhaps harder to understand how mental abilities might also have evolved in exactly the same way. We can imagine a father telling his son how to light fires but we cannot imagine the son being taught to dig through his memory banks or being told to take a step back and become self-aware.

To see how such habits of thought could be taught to each generation, we need to focus on the total social world of the child. A child may not be explicitly taught how to be self-aware, fantasize, or remember, but if it lives in a world that assumes these abilities to be present in an individual, the child will quickly pick up the expected habits of thought. For example, learning to say "I" in a sentence is a huge push toward developing a sense of identity. A mother telling a toddler to say "I want the ball" rather than "want ball" is unwittingly drawing the child's attention to the fact of its own existence and planting a seed that will eventually grow into a tangled net of ideas about a thing called "self."

Learning how to dig through our memory banks is another ability we unwittingly learn as youngsters. Toddlers live their lives in the present, as an animal does, and memories are mobilized only for recognizing and understanding all the things that parade before their eyes. Fairly rapidly, however, children develop an adultlike control over their memory banks. This is not a difficult trick for them to learn. Once they learn to speak, they are equipped with a tool that is superbly developed for controlling thought. The children of *Homo erectus*, a million years ago, would have had a struggle, using protolanguage, to master memory, but modern children have the benefit of the refinements of thousands of generations of language evolution. Moreover, children are born into a society that quite simply demands that they have control over their memories. All culture depends on it. Children find that they cannot take part in the adult world until they have picked up the knack of remembering and imagining. From the earliest age, they find they need a memory for everything, from not forgetting to wash their hands before meals to remembering the time their favorite *Spiderman* cartoon will be on TV. Children also find themselves surrounded by parents who constantly pester them to show better memories and schoolteachers trying to drum facts into their heads. Without really trying, society puts great pressure on children to hurry up and learn to control their memories.

Once children have learned to talk and started to feel the full weight of society's expectations, it is only a small step to picking up the tricks of remembering and imagining. We do not need to be explicitly taught the sort of techniques we use for remembering coffee drinking or corridor meetings. Day by day, as youngsters, we work out the necessary techniques, such as questioning ourselves with the inner voice or imagining the setting to jog our memory. Of course, the parental prompting we are given as youngsters may serve as a model for what we do later—our inner voice replacing the questioning of our parents—but generally it seems to be left to each child to discover the techniques as best it can.

More attention seems to have been paid to training memories in the earlier days of human history, when people lacked the books, newspapers, diaries, photographs, and public records of the modern world and so learned all sorts of memory tricks to help in everyday life. Early horse traders cut notches in sticks that allowed them to remember the details of hundreds of deals; medieval scholars learned their texts off by heart because so few copies of books existed; and traveling troubadours could remember new stories and tales, hundreds of lines long, after just one hearing.

These feats of memory needed hard work. One favorite technique was to study a real-life building so that the person had a bright mental picture of all its nooks and crannies. Then the person would imagine walking through this building and placing the various facts he wanted to remember in odd corners. Later he could mentally retrace his steps and methodically rediscover all the facts—a technique much like that used by Shereshevskii when he created stories to remember complex mathematical formulas.

Modern society has lessened the need for such memory. We now have plenty of books, diaries, and photographs to jog our memories. The emphasis has switched to understanding the broad principles of life rather than memorizing all the little details. With the rise of computers and videos, our external aids to memory are becoming more and more powerful, and eventually we may have all our lives on record and instant access to all the world's knowledge. We will be using machines to dig out the memories we want. The effect that this automation of our memories will have on our mental abilities is uncertain, but because of the importance of memory control to all human thought, it could be considerable.

We have just seen how we are pushed into taking control of our memories without even realizing it. Many people, however, find it hard to believe that their own minds could owe so much to the social environment they grew up in. The mind is often considered to emerge complete and intact at birth, just as the rest of our body is born with all the necessary organs and parts firmly in place. We do not see society coming along and tacking on a missing arm or kidney a few years later. Although we can accept that the mind may need to grow to maturity like the body, we find it hard to imagine that important bits of our minds might simply not have been there when we first emerged kicking and screaming from the womb.

Looking back to our earliest childhood, we probably feel that our minds unfolded gently like a budding flower. There is probably a blank period covering the first few years of our lives, for which we have no memory apart from secondhand tales told to us by our family. We tend to put the lack down to a poorly developed memory rather than to the fact that we had not been trained to form personal memories at that stage. Yet by looking at the evolutionary story and watching the stages that children go through, it is clear that we are all born as mentally naked as we are physically naked, with only natural abilities like recognition, perception, and awareness. Then, like dressing a newborn child, society comes along to clothe us with the habits of thought that create the modern human mind. Unfortunately—from the point of view of understanding ourselves—we are not in a position to see this happening. We cannot remember the days before we had learned the trick of remembering, nor can we recall what it was like before we formed a sense of self-identity, before we learned to think logically or learned to imagine. We can only watch others and hope to see how it happens in them.

Memory is only the broadest of the mental lessons that we learn from society, which teaches us almost everything that makes a human mind different from an animal's. It clothes us in all our mental habits, from broadly based abilities, such as memory control, to the fine details, such as the ideas and emotions we are likely to have. It catches us in a claustrophobic grip from the day we are born and fills our heads with the essence of thousands of generations of cultural evolution. Society is like a mold that has been carved out over a million years as each new generation of humans has inched forward in its habits of speech and thought. When humans die, the mold lives on—becoming richer and more intricately carved with each passing generation.

We could almost look on society as a vast parasitic animal that lies in wait for every new generation of minds, seizing newborn babies and stamping its self-perpetuating pawprint on their brains. To begin with, the flesh and blood of this parasite was a thin web of tribal customs, toolmaking knowledge, and hunting lore that lay over the wandering bands of early men, but the parasite hooked into men's brains through the use of words and created the mental abilities that led to the rapid rise of civilization. The infection swept man along in an explosion of population and culture, eventually causing him to set up the great social institutions of schools, churches, and courts which became the efficient modern factories for mass-producing new minds. After several thousand generations of growth and mutation, society evolved to take a tight grip on the minds and lives of every individual. Its rule is now so strict that even trivial social blunders are punished sharply, the dropping of a knife at a smart dinner party or sneezing at a wedding service being dealt with by pangs of shame as if such small errors were a matter of life and death.

Treating society as a parasite is of course an exaggeration but it helps to highlight the grip that society has on us. We live our lives so deeply buried in the embrace of our cultures that it is only when we try to swim against the tide of society that we feel its vast weight bearing down on all our actions and thoughts. Perhaps the most obvious example of how deeply rooted the influence of society can be is our sense of conscience.

Conscience feels like a very private and important part of our minds. It seems to be our innate sense of what is good and bad, and is commonly spoken of as if it were the part of us troubled by strong feelings such as guilt, shame, and, sometimes, even relief. But conscience is a good example of a habit of thought that has evolved over generations and whose pattern society now stamps on every newborn mind. It would have started as a shared body of rules and taboos developed by early communities. A band of early men would find that doing certain things proved to be sensible in the long run, such as deferring to the wisdom of elders or not committing incest. A single individual, taking a short-time view, would not be likely to see the advantage behind such rules, seeing no reason why an old hunter should be correct in guessing which way a herd of deer will run or what harm would be done in going off with his sister. However, if a large number of individuals could somehow pool all their accumulated knowledge about genetically defective offspring or the likelihood that a lifetime's experience would make the judgment of an old hunter a little sounder, the chances of the group's survival would be dramatically improved.

Language gave early man just such a way to evolve and pass on complex ideas. First, words could pin down not only specific actions but also general concepts like good, bad, honorable, and unlucky. Second, many individuals could gradually build up a rule without any single one of them having to understand just why it would work. The payoff from each act of respect to an elder might be unnoticeable to the individuals involved, but over many generations a respectful tribe might flourish and overrun its less organized neighbors.

Once socially useful ideas had become established, policing them would not be difficult in a close-knit primitive society. As seen in the way the chimp troop rounded on the troublesome young male, Dandy, who was trying to push his way to the top, even speechless chimps can maintain certain standards of social behavior by joint action. In primitive human societies, like the Amazon jungle Indians, an individual who breaks the social code soon finds out about it from the angry shouts and cuffs of his elders. Serious offenses will lead to banishment—a fate amounting to death when life depends on the support of a group. Individuals are also kept in check over the smallest of social rules with the valuable tool of ridicule: Few people can stand being laughed at or taunted for long.

With time, then, early man would have built up a baggage of rules for correct behavior. When a group saw an individual breaking the rules, the recognition of the situation would prompt an outraged reaction in their minds and they would act swiftly to bring the offender into line. At some point, this pool of knowledge in every head must have started being used internally in the way that we call conscience. Having learned to recognize wrong in the outside world and automatically give vocal warnings, man would find that he was turning this habit around on himself. If an antisocial impulse formed in his own mind, it would trigger the same alarm bells. Memories of socially correct behavior would be aroused and would issue their warnings, but this time the warnings would not be aloud to another individual but spoken through the inner voice as a warning to the self. Like an inner policeman standing duty for society, conscience would patrol our minds, ready to nip naughty ideas in the bud.

If we ignore these verbal warnings and go ahead, we are

normally hit by feelings of guilt. In some cultural settings—
like the Roman Catholic Church—people can be trained to
feel extremely strong pangs of shame and unhappiness when
they know they have broken the rules. The unpleasant pangs
of guilt come from imagining what it would be like if others
knew what we had done and were standing around telling us
off. The anxiety caused by this imaginary condemnation can
be almost as strong as that which would be caused by real
people.

Conscience can also reward. Just as guilt leads us to imagine the condemnation of others, so when we do something
we believe to be socially correct, we reward ourselves with a
mental pat on the back, imagining the praises of our fellows.
Because we have learned to punish and reward ourselves in
this way, we are turned into reliable and trustworthy members of society, behaving correctly even when we know that
no one can possibly see what we are up to. Humans can of
course ignore their consciences. We can argue with the nagging inner voice and find reasons to act differently, or we can
ease guilt with rituals like confession. But this only shows
how often the demands of society clash with the selfish urges
of the individual. We must in fact be putting society's needs
first, most of the time, otherwise society would not be flour-
ishing in the way that it is.

Conscience shows the subtlety with which society works.
Society shapes the human mind so that it is tailored to fur-
thering society's own success, yet one of the ideas society
implants in all of us is a sense of our own individuality. Ironically, it teaches us the skills of self-awareness and self-control, which are what make us feel free and independent.
Society can only afford to give us this sense of personal freedom because such independence has proved its value over
thousands of generations of cultural evolution. It is what has
made us such an inventive and adaptable race.

When we look closely, what are we using all our supposed independence and inventiveness for? Normally we are hard at work doing our best to fit neatly into the societies we live in. We may feel that we are made free by society—but only free to try our hardest to be a part of society. The truth is that we feel free only for so long as we swim with the tide; as soon as we turn and try to swim against the tide of society, we quickly feel its full weight. And we rarely even feel the desire to swim against the tide. Generations of cultural evolution would not have been wasted on forming a mold that might encourage such thoughts.

This might not seem to square with our impressions of modern society, which is commonly lamented as breaking down, portrayed as a dog-eat-dog world in which humans are becoming selfish and destructive. However, we humans are still fundamentally as much social animals as we ever were. The change is that modern society offers hundreds of different and often contradictory life-styles to which we can conform, and the cracks that show in society are essentially cracks between subcultures rather than between individuals.

When people lived in simple, primitive societies, they really had only one possible life-style. All their customs and ideas would have evolved as a package and remained unchanged for hundreds and even thousands of years. Individuals were unlikely to rebel against their societies because they literally had no idea of a different kind of life. Moreover, their life-style would have been perfected over many generations and so would be one of harmony. Indeed, researchers who have had contact with the world's few remaining tribes of primitive people remark on their contentment with their way of life. Even the slightest contact with modern man, however, brings with it alien ideas that can destroy within a generation the carefully evolved balance of the primitive tribe.

Early man probably lived the same sort of contented life.

There would of course have been deaths, illnesses, and jealousies, but every mind in a group would have shared the same childhood training and learned the same adult lessons. Like conscience, these ideas would be absorbed as a net of knowledge and could be expressed to the self through the inner voice, but because early man had only one neatly packaged set of ideas, he would not find any cause to argue with himself in his head. He would have only one point of view—that of his tribe—so would lead a life of unquestioning acceptance of the ideas he found in his head.

Modern man clearly lives in an entirely different world. While primitive societies are dominated by the weight of their traditions, modern Western man faces a confusing variety of traditions. He might belong to such vastly different subcultures as a religious order or a street gang; he might be a soldier or a dropout, a factory worker or a doctor. There is not just one way of thinking and acting that is considered right, but many.

Even when modern people adopt a clearly defined subculture to live in, they can still find themselves facing many contrasting social situations during the day. The average worker passes through three in an hour every morning—leaving home, passing through the rush hour, and walking into the office—each of which has its own ethics, ways of behaving, and likely patterns of thought. To cope with this, the modern mind has to be able consciously to shift gears. When people arrive home from work at the end of the day, they may have to drop the stony face that might have been appropriate while in a crowd of commuters and shift into a family frame of mind. Primitive man may have been deeply and unquestioningly embedded in his one culture but modern man bounces from work to club, from home to public places. He needs a general background level of control over his frame of mind if he is to put on a social front that is right for the occasion.

Modern man is so skilled at presenting the right face to go with the situation that it is almost as if he has a bagful of different masks. We learn to hide our true feelings behind a front and play a role that fits in with the expectations of the people around us, so the sales clerk smiles helpfully through gritted teeth at the dithering shopper and the partygoer laughs along enthusiastically when bored by the company. Just as we put on the correct clothes to go to work, to a party, or to play tennis, so we put on the appropriate social face to go with the occasion. We find it normal to mask our inner feelings and act the part dictated by the situation. The only time we might let the mask slip is when we lose control through extreme emotion or drunkenness.

Ideally, we like to feel completely "at one" with the occasion. We are happiest when our innermost feelings and thoughts mirror the social act we are putting on, so we enjoy being swept along by the warmth of friends as we share coffee after dinner or the anger of a group talking over some injustice. We relish the shared scorn of gossips talking about someone else's behavior or the formal pomp of a town council meeting. But even when we cannot manage to tune ourselves in as completely as this, we would rather try to fake it than disrupt what is going on.

The way we all use social masks is a sign of the pressure we feel to live the sort of lives dictated by society. Each of us is like a piece in a giant jigsaw puzzle and we know that if we do not adopt the right social shape, the jigsaw of society will fall apart. We have been trained from an early age to do our best to fit in—and to want to do so. It is part of the mold society has evolved that we all have an in-built urge to belong, even the apparently antisocial sections of society. Hell's Angels, punks, and hippies, for example, are supposed to be reacting against society but in fact they are defining their own subcultures by exaggerating its contrasts with the broader mix of society. To maintain their group identity, members actually have to keep up an even stricter social mask than usual—punks have to be cynical, Hell's Angels aggressive, and hippies happy and tranquil. Humans naturally feel a broader range of emotions than this, but the need to preserve the group's stereotype means that individuals are even more guarded than normal in the feelings they give expression to. The sacrifice is usually willingly made for the pleasure and security gained in being part of a close-knit group.

We have, then, seen how man created society and how society has in turn shaped man. Early man lived in small pockets of society which were only large enough to sustain one way of life, but modern man has brought hundreds of types of societies crashing together. Cosmopolitan cities have thrown different religions, cultures, races, and politics into a melting pot of ideas, out of which has come wave upon wave of new cults, fads, and fashions as the rich mix of ingredients combines in new ways. The richness of Western society gives man a new freedom to choose how he wants to think and act but it has also become difficult for modern man to feel fully at home in his society—though he has the deep desire implanted in him to try. The best he can do is break his personality into pieces and learn to put on the right mask for whatever social situation he finds himself in.

If we learn to put on a convincing enough mask, we can create our own little pockets of society and rediscover the warm camaraderie of our ancestors seated around the stone-age campfire. But because we are conscious and aware of our individuality, we have constantly to guard our minds if we are to catch the impulses and utterances that might spoil the front we are putting on; our attempts to melt into the rich complexity of modern society serve only to make us far more self-aware than our ancestors. Our efforts to fit in highlight our separateness.

This leads us to the possibility that self-awareness could

be a side effect of all the skills that society teaches us. Society obviously benefits directly from the extensions such as speech, conscience, rational thought, and memory control, which it grafts on to our impressionable minds. We would therefore expect that the evolutionary pressure to improve these skills would be strong and that they would become about as polished as cultural evolution could make them. However, our sense of self-awareness might be a different matter.

Early man could obviously have become self-aware quite soon after language arose, for memory control would have made it possible for him to step back and think about his own thoughts and the fact of his own existence. But he might have spent very little time doing so if society found no particular reason to encourage such behavior. Also, the skill with which he analyzed what he saw might not be very great if his mental tools had not been honed to a sharp edge by the pressures of evolution. Self-awareness might have been a rather haphazard and poorly focused skill, if it existed at all in *Homo erectus* about a million years ago.

Modern man may have recently sharpened his sense of self-awareness through scientific understanding and the need to put on social masks, but self-awareness is still not likely to be at the top of society's list of priorities. It may have only an indirect benefit—or even work actively against a harmonious society—so it will not have felt the full shaping force of cultural evolution. We may not therefore be nearly as skillful at self-awareness as we could potentially be. Indeed, with a little extra effort, we might be able to improve on what cultural evolution has given us. We will look in more detail at self-awareness after finishing the story of language and thought.

We saw how early man progressed gradually from slow protolanguage to rapid internalized speech. Then the explosive mixture of modern language and the easing of the ice

ages sent *Homo sapiens* rocketing upward on an evolutionary spiral of civilization and mind powers. Man's speech-
sharpened mind could take quick advantage of the better
climate and within a few thousand years man was in the
middle of a population and culture boom that made him the
dominant species of the planet. However, language alone was
not enough to fuel the dramatic rise of civilizations all around
the globe. The real advances in culture came with the invention of a permanent form of speech—writing.

Both writing and mathematics existed at the dawn of civilization. As soon as man settled in villages supported by
farming and trading, he needed some crude form of docu-
mentation, and early civilizations came up with a variety of
ways of counting and recording. One of the earliest methods,
invented in the Middle East, was to seal stones in a ball of
clay, with each stone standing for a certain amount that was
being recorded. Then, so people would know what was inside
without having to smash the ball open, they started to score
marks on the outside. Soon they abandoned the stones alto-
gether and simply scratched on flat tablets of clay.

This shows the amount of luck involved in the story of
language and the rise of humans. Other civilizations, such as
the South American Indians, used beads and knots on pieces
of string, which may have served the original purpose well
but did not have the lucky features of clay that led to the
first marks of writing. At first, both writing and counting
were done with picture representations rather than real symbols. Five baskets of barley would be depicted as five scratches
alongside the outline of a basket. It took a few thousand years
for people to start using pure symbols such as numbers and
letters, and indeed some cultures never broke away fully from
picture systems. The Romans, for example, used a sophisticated version of scratches for counting—the familiar Roman
numerals—which meant that they did not get much farther

than simple arithmetic, handicapped by the lack of such things as decimal places and negative numbers. The Chinese stayed with pictograms for writing, which meant learning forty thousand symbols (compared to the twenty-six letters of the English alphabet) to become literate and consequently helped limit reading and writing to a small elite.

The Western alphabet can be traced back to the Egyptians. They also used pictograms, but the pictures stood for sounds— as the letters of the English alphabet do—as well as for the things they were the images of. The Egyptian system was picked up by passing traders, who simplified it and spread it around the Mediterranean. Eventually the Egyptian pictures were stripped down to pure symbols of sound, reaching ancient Greece in the time of Plato to provide the basis of the modern alphabet. Further advances, such as pens and printing presses, relieved the slow tedium of writing on wax tablets and leather parchment, and the flood of books from the printing presses gradually led to standard forms of spelling, punctuation, and grammar.

Another major advance was the internalization of reading and writing. Just as early man must have first learned to speak and later got into the habit of the inner voice, early readers and writers first had to perform their skills out loud. The monks of the Middle Ages, who hand-copied Bibles, whispered the words to themselves as they scratched away with their quills and most readers of those times had to mouth written words out loud to take in their meaning. The few scholars who had the skill to read silently were treated with considerable awe and suspicion.

In modern times children are painstakingly taught to read in their heads. Some of them may continue to make slight movements in their throats even though no sound is produced, which slows their reading down. All the fastest readers learn to translate direct from the page, the words going straight from eye to understanding and bypassing the vocal equipment. Silent reading has great advantages, in particular, that sentences can be read far more quickly than they can be spoken; a reader can thus pack his limited working memory with a richer mix of ideas than can a mere listener.

The story of writing is one of rapid technological improvement, going from a slow, inefficient tool of a few to a fast, efficient medium for the masses. However, the real impact that writing had on civilization was that it became a new form of genetic material to capture the advances of human culture. Just as spoken language had earlier helped man to pin down each generation's advances in food gathering, tool-making, and social living, so writing became the genes for capturing the rapid strides being made by civilization. It provided a permanent way of recording the details that underpinned complex societies and that were too numerous or boring for the human mind to remember easily—such as taxation records and business deals. As writing became a more fluent tool, thinkers could use it to preserve the minute detail of their discoveries and knowledge. Each following generation would then build atop a foundation of ideas rather than having to discover everything afresh for itself.

We take this for granted today with all of our magazines, newspapers, libraries, and colleges—sometimes it even seems too much is being put into print and reverently kept for posterity—but most of the discoveries of the first great civilizations would have been lost after the collapse of the Roman Empire without the chance survival of a few hundred books. The chaos of Europe's Dark Ages meant that much of the ancient world's knowledge was forgotten, and if not for the preservation of a few Greek and Roman texts, which became the basis of a European revival in thought and science, today's technological society might not have arrived for several centuries yet—if at all.

Writing not only came to serve as a warehouse for knowledge but also led to more advanced forms of thought. The act of writing removed the emotional heat of face-to-face debate, so thinkers could approach their work with cooler heads. Also, because a sentence had to be grammatical and complete to be understood by the reader, writers were forced to make explicit the assumptions of their arguments, and once their ideas were down in cold print, it was easier to go back and check for flaws of fact or logic.

So far we have been talking about how writing has helped the advance of higher thought, but writing had a much broader impact than this on civilization because of the way it captured ideas and helped form the mold of society that shapes our minds. Until writing came along, man had only spoken language with which to do the job of training the next generation to think. This was, of course, enough to create the habits of memory control, higher emotion, and self-awareness, but parents and tribal elders could remember only a limited amount of knowledge themselves and so were handicapped in what they could pass on to their children. However, the written word could capture the fine detail that people could never remember, making all of a civilization's experiences and knowledge available to the individual in a handy prepackaged form.

Nowadays, once a child has been taught the basics of language and memory control through the direct teaching of those around it, the rest of its mind tends to be molded by the rich world of the printed word. Many ideas about, for example, love, fashion, or politics, are too complex or emotional to be talked about face-to-face and most people learn about them from novels, newspapers, magazines, political pamphlets, and so on. Over the past five hundred years the printed word has come to form part of the very fabric of our societies, with the minds of each new generation shaped as much by attitudes discovered in comic books and trashy novels as by the stern moralizing of parents. Recently the rapid advance of technology has led to the printed word's being heavily supplemented by the visual advances of photography, TV, and film. These can be even more powerful media for passing on cultural ideas, and computer technologies are now promising a further breakthrough in the use of culture to shape minds.

One of the problems of modern science is that while research papers are constantly being produced, it is almost impossible for researchers to find all the papers relevant to their line of work. It is said to be cheaper and easier to rediscover something than to dredge it out of an academic library system. Computers, however, will soon be able to search the mountains of information that humans produce and make sure that a worker has every possible reference at his fingertips. Computers will act as super-memories that can both store and retrieve all of recorded human thought. And it is not just scientists who will be able to use these super-memories, for the falling cost of computer technology will make them available to everybody. Computers will also revolutionize communications, making it possible to search the country for like-minded people to talk with, to form electronic clubs for sharing ideas, and for cheaply publishing a person's own thoughts.

All this will lead to more of our culture being put "on the record." People will be able to set up electronic filters to sift the computerized world of information for everything that fits their particular interests. They will also be able to get in touch with enough like-minded people to form groups large enough to evolve their own ideas. Western society is already splitting into subcults and special-interest groups with their own ways of thought and ways of living. Computers will speed this splintering of society. Perhaps these new pockets of culture will then create useful new ways of thought, which will eventually be fed back into the mainstream to help advance society as a whole.

So far technology has mainly helped man physically. We can grow more food, travel farther and faster, cure diseases, and live in comfort. But now technology is starting to open the door to rapid mental evolution. Technology will automate the process of fixing cultural advances and weaving them into a social mold that shapes the minds of the next generation of humans. This new technology might of course end up being mostly used to search out trivia or publish half-baked rubbish. But scattered among the dross could be hidden many gems. The secret would be to set up electronic filters that sift the world for only the most interesting ideas. And since our higher emotions, opinions, and habits of thought are culturally shaped by the people we surround ourselves with, we could seek out the like-minded to work toward the sort of minds we would ideally like to have.

Leaving such speculation aside and returning to the story of how our minds were created, we have focused so far on what might be called our window on the outer world. We have talked about the distilling of sights, sounds, tastes, and touches that lead to a conscious awareness of our surroundings. We have seen how these sensations are preserved as traces in memory and then deliberately reawakened with words to form imagined scenes and thoughts. But the brain also has its window on the inner world of the body through the emotions. These are also part of our consciousness and this inner window has proved an equally fertile building ground for the sort of mental extensions that are created by cultural evolution and language.

SEVEN

Pure Emotions and Romantic Notions

Life was much simpler when the first creatures appeared on the earth about 700 million years ago. In the warm primeval oceans that covered the planet, worm- and jellyfish-like animals drifted in the currents or wriggled in the bottom ooze. Because of the lack of predators, these ancestral animals could afford to be little more than soft-bodied eating machines, living to the tune of inner rhythms, blind to the world around them, and more or less bumping into the bacteria and algae that were their food. There would have been some rather crude sensing of the world outside through simple reflexes. A few light- and touch-sensing circuits—much like those of the sea slug, *Aplysia*—would have been enough to keep these animals moving along in the untroubled ocean shallows. Their lives would have been dominated by eating and breeding, activities kept under control by slow-changing internal hormone cycles.

This state of affairs changed after a few hundred million years when more complex animals, like squid and fish, started to evolve. These were hunters with keen senses and powerful muscular movements which allowed them to chase their dinners. At this point, evolution had to come up with the brain. Animals could no longer be watery bags of guts and sex organs, controlled by a clutter of hormone cycles and simple reflexes. They needed a central organizer that could draw everything together—both the needs of the inner body and the threats of the outside world—and make intelligent decisions about what to do next.

From the beginning, the brain was designed to look both ways: outward to form a picture of the world around an animal and inward to sense the inner needs of the body. From looking to the outside, the brain would get information about potential threats and possible meals; from the inside, the brain would pick up signals that the body was hungry, thirsty, or tired. The brain would also get feedback on the effect that brain-directed actions were having on the body—such as whether a certain food tasted good when chewed or whether pushing through prickly undergrowth was damaging the skin. The brain could then weigh up these inner urgings against the outside picture to set an animal on the best course of action. A jellyfish is a fairly brainless and undirected creature which swims along, continuing to feed, ignorant of the fish jaws about to close around it, but an advanced animal like a rabbit, sitting chewing grass in a field, will ignore its hunger pangs and race for its burrow on sighting a fox.

In the animal kingdom, mammals such as the rabbit are the most emotionally complex of creatures, with the richest variety and the finest shades of emotions. The reason is that mammals are also the animals that have the greatest control over their internal states. They leaped ahead of their reptilian ancestors when they became hot-blooded. The early mammals could not only raise their body temperature by burning more fuel but also found other ways to increase the metabolic activity of their bodies. They developed a muscular diaphragm and rib cage so that they could suck more oxygen into their lungs and evolved a better-designed heart, which could pump more blood around the body. In these and many other ways, mammals changed so that they could react more energetically in a crisis.

As we have seen, evolution does not encourage waste. The only energy that counts is the energy devoted to breeding and passing on genes, so despite their superbly athletic bodies, mammals could never afford to live their whole lives at full tilt. They would simply burn themselves out. But by having control of their moods, mammals could strike an economic balance between relaxation and excitement. They evolved a rich array of emotions as their controlling mechanism to lower the body's thermostat when the environment was safe and to hit the panic button the instant that danger arose.

Humans share this ragbag of drives, passions, body feelings, moods, and urges that we lump together under the general heading of emotion. But on top of these, we also have distinctively human emotions—the "higher" emotions such as love, patriotism, humor, and guilt. To try to nail down all the possible forms that higher emotions can take involves a welter of labels, for ideas like happiness or anger can be divided up into hundreds of shades of meaning. As well as happiness there are ecstasy, rapture, elation, joy, contentment, and so on; anger can be rage, fury, cold annoyance, or disgust. Significantly, there seem to be as many shades of higher emotions as we can think up words to label them.

Disgust, for example, can be subdivided into the kind of disgust we feel with smelly objects and the kind we feel with unpleasant people. We can go on to draw out the differences that characterize the two shades of disgust: One is more physical—hitting us in the nostrils and stomach—while the other is perhaps more cerebral. We can continue that there is disgust for morally unpleasant people and another type of disgust for inadequate people, one of which is tinged with anger and the other perhaps mixed with pity. And so it goes on until we have divided an apparently basic emotion like disgust into dozens of shades of emotion.

Psychologists get little further when they try to fit the many shades of emotion into a few broad categories. One scheme divides emotion into four types—level of arousal, feelings toward the environment, general moods, and feelings about one's own competence But such schemes still do not find a proper place for many of the more complex emotions such as artistic enjoyment, guilt, vanity, humor, and honor.

The problem is that higher emotions are human inventions. Through the creative force of cultural evolution and the building material of language, we have built artificial extensions to our natural emotions. Attempts to categorize higher emotions are doomed because the very act of labeling is tantamount to inventing new emotions. Dividing up disgust into thing-disgust and moral-disgust creates new tags around which we can hang more narrowly defined nets of ideas. A cake will eventually crumble if we cut it into finer and finer slices but words are solid symbols and can be handled just as easily by the mind whether they carry a heavy load of ideas or the barest covering.

We could invent a new emotion right now—aardvark-disgust, for instance, which is a disgust with the South African animal with a motheaten tail, rough bristly nose, and slightly comical, floppy ears. It is rather an ugly, pathetic creation, but perhaps also slightly amusing and having the strange honor of being placed first in the dictionary. Even after this brief summary, the word *aardvark-disgust* should conjure up a particular mix of emotions in the hearer's mind. It will take its place in our vocabulary if we find it useful and be dropped into conversations as easily as any other word with the grandest weight of ideas behind it. *Aardvark-disgust* shows how complicated the story of the higher emotions can be and we should first identify the raw feelings that can rise up into the conscious plane before looking at the way cultural evolution has built extensions out of language to give us our complex human emotions.

As has been said, humans inherited a ragbag of emotions from their mammalian ancestors, and like chimps and monkeys, they are naturally volatile and excitable. However, if we sift through the jumble of emotions and try to pick out the purest unadorned forms, we can boil emotion down to the three basic elements of arousal, pleasure, and pain.

Arousal describes the level of energy coursing through the body or, to put it another way, the level of the body's thermostat. It is largely controlled by the primitive bulge of the brain stem—the bit of the brain that plugs into the ball of the wrinkled cortex like the stalk of a mushroom. The body's arousal levels can range from a lazy drowsiness to extreme heart-pumping excitement, according to whether we are, say, sitting in the sun on a quiet Sunday morning or swooping down the highest slope of a roller coaster.

The arousal center in the brain stem has a direct chemical effect on the body. The center pumps out certain hormones to quiet the body down and others to kick-start it into action. After a sudden shock the heartbeat speeds up to pump more blood around the body, and we take deeper breaths to oxygenate the blood, squeeze extra red blood cells out of the spleen, release sugars from the liver, cut down the blood supply to nonessential areas like the skin and gut, widen the pupils of the eyes, and raise the level of wound-repairing agents in the blood. Relaxation signals bring the body back down to the resting state, so they mostly have the exact opposite effect on the body's organs. The messages put out by the brain stem depend on what the brain is seeing of the outside world. The conscious cortex keeps an eye out for danger or excitement and when it spots something happening, it flashes an alert to the brain stem to get the body pumping. Like a general passing orders to his chief of staff, the higher brain does the thinking and gets the brain stem to rally the troops.

The rapid panic responses of the brain stem are only one of several mechanisms with a controlling hand on the body's metabolic thermostat. We also have long-drawn-out cycles and biorhythms, which some days leave us in a generally good mood with plenty of energy, while on other days we are left tired and down. Such underlying long-term patterns of arousal are affected by many things, such as our fitness, our sleep, and our levels of stress.

Whether we are experiencing immediate or long-term feelings of excitement and relaxation, an important question is how these feelings enter our conscious mind. From what has been said before, these feelings would need to become one of the firing nets of cells in the cortex that make up our conscious plane before we could experience them. So far, we have talked about the cortex analyzing the outside world and triggering the brain stem, but we have not accounted for the very real feelings that then accompany the body's reaction. The answer is simple: The way we consciously experience different levels of arousal is through noticing the many small reactions that our bodies make. Our conscious experience of relaxation comes from telltale sensations such as loose muscles, a quiet heart, and settled guts, while our awareness of excitement is through the pounding of the heart, a sinking feeling in the stomach, and a twitchiness of the body. These small sensations are the firing nets that enter our conscious awareness. As well as these rather indirect signs, relaxation and excitement are also felt through their direct metabolic effect on the brain. Arousal signals send oxygen and blood pumping to the brain just as they do the rest of the body, so we cannot help but notice the way our attention then becomes focused and razor-sharp, our thoughts rapid and clear. By contrast, relaxation brings with it a blurring and dimming of consciousness as our minds drift quietly in neutral, saving energy on a low thermostat setting.

This spectrum of energy levels that we experience, ranging from relaxation to excitement, is about the nearest we can find to a "pure" emotion. But it is very rare for us to think about emotion in such unadorned terms. Usually we talk about the passions—feelings like fear, rage, lust, and aggression—which are based on the foundation of arousal. They are labeled differently according to the likely behavior that goes with them, although fear and rage, for instance, are based on such a similar body reaction that they are really two sides of the same coin. When we say we fear something, we mean that we have experienced something that makes us excited and likely to run away; the emotion we term fear is a mixture of high arousal and a typical pattern of action. In the same way, rage is the triggering of identical arousal pathways but with the likelihood of quite a different reaction—attack.

At first glance, this view of fear and rage goes against what we think we feel. Because of the perception-sharpening nature of words, we tend to treat fear and rage as two clearly distinct passions. When we talk about fear, we think about the arousal and desire to flee as an inseparable package, quite different from rage's arousal and desire to attack, but fear can swiftly turn into rage when we are prevented from fleeing and so forced to go on the attack. If you suddenly saw a snake slithering across the floor of your living room, it might strike fear into your heart and you would shrink back in your chair with nerves jangling, ready to leap up and run. Then you realize that you must do something about the snake before it gets lost in the furniture and, shoe in hand, you run forward and hit it on the head. After the initial fear, you are now dancing around, pounding the snake in blind fury until it is long past dead. The label that you would attach to your extreme excitement would depend very much on your actions.

Other passions—such as lust and what we might call cold-blooded aggression—are also based on the fundamental emotion of arousal. They too are flavored by the circumstances in which they arise and the behavior that follows. However, there are some subtle differences in the body's physical responses with these other passions. Lust pushes the metabolic panic button just like fear and rage, but the sexual response involves not only a pumping heart but also particular body changes, such as an erection in the male. In cold-blooded attacks—the kind of calm aggression shown by cats and other carnivores in bringing down prey—the body is tuned to a clearheaded and unwavering attack and under these conditions the brain stem pumps out a slightly different mix of messages so that the body is aroused but not frantic. The brain stems of the big cats secrete a much richer mix of noradrenaline—one of the hormones associated with calm attack—than do those of apes, for example. On the other hand, apes are more likely to pump out adrenaline, the hormone associated with jumpy defensive attacks of rage.

These variations in the emotional makeup of the big cats and the apes are of course wired in by evolution. Tigers and leopards are big predators that chase and kill to eat, so they need a powerful but clearheaded surge of energy. Apes are highly social and largely vegetarian animals, with little need for this calm aggression; if they attack, it is usually from defensive fear. Humans can hunt game and play sports with the same sort of cold-blooded aggression shown by cats—our recent hominid history as hunters might even have strength-

ened this sort of metabolic reaction in us—and we can also
make what are natural rage-type attacks on people and things
around us if escape or avoidance looks impossible. Humans,
however, have stepped over the natural boundaries and learned
to use cold-blooded hunting aggression in social situations.
We have developed the cultural training methods that can
turn people into soldiers or street fighters and so twist natural abilities to new ends.

This sort of excited but controlled passion leads to people
who have risked danger being often asked by onlookers if
they were not afraid. The heroes of the moment usually give
the socially acceptable answer of yes, everyone feels fear. But
fear is in fact how arousal feels to passive bystanders. The
active firefighter feels a controlled excitement and may feel
fear only when he later passively remembers the dangers he
went through; hence the typical response: "I was too busy to
be frightened and it didn't hit me until later."

The importance of context to the way we think we feel
has been graphically shown by a simple experiment in which
people were injected with adrenaline to make them aroused.
If nothing was said about the drug, the subjects reported the
plain physical effects on their bodies such as a dry mouth,
trembling limbs, and a rapidly beating heart. But when told
that the drug would make them happy, they felt happy and
when told the drug would make them angry or fearful, then
they felt those emotions instead. The subjects did not simply
mistake their arousal for the emotions they had been led to
expect but genuinely experienced these feelings. The mixture of arousal feelings and thoughts about what the arousal
feelings might mean are all we need for the full experience
of an emotion.

Having looked at the basic relaxation/excitement spectrum, we can try to isolate what on first glance looks like a
second spectrum of pure emotion—the pleasure/pain contin-

uum. However, while pleasure and pain are opposite in their likely effect on our behavior, they are not like arousal in belonging to either end of a smooth gradient of the same emotion. Rather, each is a feeling in its own right with its own center in the brain stem and each has its own spectrum of arousal levels. Pleasure can range from gentle contentment to rapturous enjoyment while pain can vary from mild unease to unbearable agony.

The separateness also means that the pleasure and pain centers can fire independently. This creates the mixed emotions that we sometimes feel are a combination of both pleasure and pain—feelings such as the bittersweet pangs of unhappy love. Mixed feelings are only one complication in talking about pleasure and pain. The two emotions are so tangled up in our higher emotions that it is very hard to see their roots in the mind, and to get a clearer idea of them, we need to look at their evolutionary background.

Pain comes in many guises, from the piercing agony of the dentist's drill to the dull ache of a stiff shoulder, but the common thread behind pain is that it is always an alarm bell telling us that our body is being damaged in some way. The damage might be anything—burning from hot water, the tiredness of overworked muscles, an acid mouthful of lemon juice, or the stubbing of a toe on the pavement. When the body's network of sensory cells spots signs of damaging pressure, temperature, or chemicals, it sends a loud warning signal clanging through the brain, which demands immediate attention from the conscious cortex.

In primitive animals, such as a shrimp or worm, their nervous systems are too simple for much more than a reflex reaction. The jangling of sensory cells automatically sparks a spasm in the tail muscles to jerk the animal away from danger. Agitation rather than pain is perhaps a better word for what they experience. In humans—and mammals generally—the jangling nerves do not travel straight to the muscles but first deliver their messages to the conscious theater of the mind. There, the warning bells rouse a fleeting pattern of cells and a pain becomes part of conscious experience. Unlike arousal, which is experienced indirectly through its effect on the body, pain does indeed enter consciousness directly. The body's senses act as a giant damage-detecting sense organ, which reports straight back to a special "pain mapping" center in the brain stem. Electrical stimulation of this center by an electrode causes feelings of pain just as if the center were receiving distress signals from the body.

In a similar fashion, pleasure probably started out as a reflex alarm bell and ended up as a rich conscious experience under the control of its own center in the brain stem. The common thread in pleasurable experiences is that they signal that something good is being done to the body, that is, good in the biological sense—such as the satisfaction of needs like eating, drinking, and sex.

Pleasure and pain thus had their roots in acts of basic biological importance such as breeding, feeding, and self-preservation. These needs led to the creation of simple reflex pathways in primitive animals like shrimps, but the pathways were like motorways from sensory cell to muscle cell so action was automatically taken without any pause to savor the feelings we call pleasure or pain. As animals built up inner pictures of the world and could take more intelligent decisions, evolution developed the parts of the brain that these reflex pathways ran through. The patches of brain became emotion centers capable of creating the firing nets that we experience as pangs of hurt or happiness.

Once this happened, evolution had the mechanism to create an inner driver for higher animals. Feelings of pleasure and pain became the general way of telling an animal whether a particular action was good or bad for it. When this judgment was coupled with memory, higher animals could learn from experience. A rat in an experimenter's box will recognize a lever on the wall and know from earlier sessions whether pushing the bar will lead to a reward of food or a punishing electric shock. Seen in net terms, the rats forge links between memory nets and nets in the pleasure or pain centers, so recognition of the lever creates a visual net that then sends a jangling message to tweak the appropriate net into life in the emotion centers. Depending on whether experience has associated the lever with pleasure or pain, the rat will become consciously aware that the lever is either something nice or nasty. Inner feelings that might have originally flagged very specific sensations in lower animals—such as painful damage to the body or pleasurable sexual satisfaction—have become general feelings of judgment in the higher animals. A mammal no longer needs to walk into trouble before it is hit by painful warnings. It can attach memories of pain to memories of dangerous situations and so be reminded to avoid them.

Although we have talked about pleasure and pain in very simple terms, as if they were always the sharply defined pangs we get when jabbing a pin in our finger or scratching an itchy leg, the pleasure and pain centers are actually more complex in their behavior. As well as producing sharp pangs like sudden electrical firestorms racing across the cortex, they play a part in longer-term moods such as happiness and depression.

The brain is both a chemical and an electrical organ—electrical for circuits that need to be fast-acting and chemical when things move slowly. While sharp pangs seem to be the result of a powerful electrical jolt to the pleasure and pain centers, moods are produced by the slow release of brain chemicals that have quite specific effects on various pathways in the brain. For example, running is a painful and tiring activity, but when an athlete goes on a ten-mile run, the

brain soon realizes that it is unhelpful to keep nagging the
runner with pain messages and instead starts pumping out
an opiatelike substance that blocks the pain and gives the
mental lift that we call a runner's high. The brain has an
armory of such natural drugs. It is because the brain makes
such a rich use of mood-controlling drugs that many plant-
derived chemicals, like cocaine and opium, have such a potent effect. The action of these mood drugs is not yet fully
understood, but presumably they stimulate or block firing in
the emotion centers and so create the conscious sensations
of pleasure and pain that flavor our different moods.

The body in fact has a number of chemically controlled
drives and urges that can make their demands felt in the conscious brain. For example, a thirst mechanism monitors the
level of water in our body and when we start to get dehy-
drated, it causes a special messenger hormone to be released
in the brain which triggers the conscious sensation we describe as thirst. Usually this is a faint itching on the fringes
of consciousness, just enough to provoke the conscious mind
to form a plan to solve the problem—to get up and drink a
glass of water—but when thirst builds up—if we are lost in
the desert, say—the sensation can start to become painful.
And when we finally get a drink, the conscious mind is re-
warded with a flush of pleasure from the pleasure center.

The mind has other chemical circuits like this for hunger
and the other body drives. At a low level, they send quite
specific messages to the brain about what the body wants.
At a higher level of intensity, they seem to start recruiting
the help of the pleasure and pain centers to make sure that
the mind pays attention. The pleasure and pain centers may
have originally been formed as part of rapid reflex loops for
damage detection or to drive sexual behavior but once they
existed in the brain, they provided the body with a pathway
to the conscious world where it could make its demands felt.

In trying to identify the purest forms of emotion in the human mind, we have so far talked of a spectrum of arousal levels that perhaps we notice only because we have got into the habit of watching ourselves and labeling such things. We notice the increase in our heartbeat, the churning in our stomachs, the feelings of sudden nervous energy, and the sharpening of our awareness, and we label these changes according to the context, and say that we feel emotions like anger, fear, passion, or relaxation. We also have a variety of good and bad feelings that come from the rousing of the pleasure and pain centers. These feelings can be felt as razor-sharp pangs at their purest and most intense, or they can be tied in with the more diffuse workings of chemically driven moods. These two spectrums of emotion can be mixed to form some of the more complex emotions we feel. Hate, for example, mixes feelings of passion and pain as we focus on someone or something in the outside world. Contentment is relaxation flavored with a low chemical glow of pleasure. Given our ability to create labels—as seen with the creation of the brand-new emotion of aardvark-disgust—we can mix these basic ingredients in different proportions to come up with hundreds of shades of emotion. The perception-sharpening power of language makes each of these emotional cocktails seem quite distinct.

Humans have been very creative in mixing new cocktails. We not only use the raw natural ingredients of pure emotion but also add a heady mixture of ideas. Indeed, some of our higher emotions—such as love, contempt, patriotism, honor, and jealousy—consist mainly of ideas. But there is nothing arbitrary about them, as there was with our invented emotion of disgust for aardvarks. Higher emotions have been manufactured by society to push individuals into behaving in the way society wants.

Before language, early man must have had a range of emotions similar to that of chimps. Like them, he probably spent most of the day quietly foraging or resting in the shade, but again like them, he would also have been highly excitable. When caught up in social disputes, he would have shown the same wild swings in temper from angry threats and screeching fear to passionate hugs of reconciliation and gentle grooming. This basic impulsiveness would not have been a good foundation for the great social advances like food sharing and coordinated action. Prelanguage man's selfish impulse would be to eat everything that he gathered and even snatch what food he could from weaker group members. He would also quickly get bored and abandon campsite chores like collecting firewood or trekking many miles to find the flint for making tools. But once language came along, cultural evolution could get to work on man's mind.

This new type of evolution created not only the thought habits that gave man a far greater understanding of the world—habits like self-awareness, personal memory, and imagination—but also fashioned new emotions to steer man along a higher social path. Biological evolution created the pleasure and pain centers as inner drivers so that an animal could balance the needs of its body against the threats of the environment. Cultural evolution then created new social emotions which forced man to weigh up the needs of his group as well as his own needs. Because of these socially shaped emotions, a man with a full stomach might well risk a lurking leopard to find food for a lame and hungry comrade. His selfish fears would be outweighed by the socially evolved emotion of loyalty or compassion.

The way we learn these social emotions is simple. During childhood, we pick up a host of ideas about socially useful attitudes. We are then trained to link this knowledge with our pleasure and pain centers. In a process little different from a rat learning whether pushing a lever is nice or nasty, we are taught what sort of social actions should feel good or bad. For example, we talk of team spirit, loyalty, and patriotism as if they are different emotions welling up from the noble recesses of our soul. But all are based on the same general net of ideas taught to us during childhood and adolescence. This general class of feeling has common elements such as the idea of belonging to a group, putting the group's needs before your own, and not letting your sympathies stray to a rival group.

Clearly this is a vein of socially useful attitudes that would help a group survive. The many different labels we use to describe this net exist only to emphasize a particular point such as the object to which the feelings are being directed. In the case of team spirit, this would be a team, with loyalty, an individual, and with patriotism, a country. At root would be a common net of ideas built up by the blurring of thousands of childhood experiences. Society teaches us the same basic lessons in numerous different ways, such as when we are put into rival teams at school, when we overhear adult remarks about damned foreigners, when we read stirring tales of patriotism, or when we feel firsthand the effects of telling tales on our schoolmates. Gradually, these specific memories build up into a rich blur of knowledge, like our nets of knowledge about picnics or car crashes, and once we have absorbed the general net of ideas about group loyalty, we find it often advising against selfish impulses. When we feel like skipping football practice or relaying gossip about a friend, we suddenly recognize that such thoughts do not match our ideas on group loyalty. We hear the inner voice that we call conscience telling us what we really ought to be doing.

However, dry advice on socially correct behavior would hardly count for much against our selfish urges, even when delivered through our ever-nagging inner voice, unless we also had painful or pleasurable feelings that tell us what to do.

Cultural evolution's trick is to tap into the pleasure and pain centers to add emotional weight to social ideas. Links are forged between the net of social ideas and the emotion centers so that ignoring the social advice will trigger pangs of unhappiness and obeying will lead to good feelings.

The teaching of these links is obvious when we watch the way parents, relatives, and society combine to give youngsters lessons on how to behave. Even before a baby can speak, a mother is cooing "Who's a good boy, then?" and tickling it playfully when he does something she wants or frowning when he does something she does not like. These small signs are very effective, since the mother's cooing and smiling are naturally pleasurable experiences for a child, while a frowning face is frightening and painful. By using these small rewards and punishments, a mother can control what the baby is feeling and so teach it to associate correct behavior with nice feelings.

At first—as with language—the mother has to do everything. She prompts the action and then follows it up with the rewarding or punishing feeling. She makes the child burp and then smiles and tickles him to make him feel happy. Or she places the child's hand on the cat's head and if the child starts to bash the cat in the face, she gently restrains it, frowns, and says "Naughty." After many years of such training, a child will have formed its own general memories of what actions are socially either good or bad. The child will also have internalized the link with the emotion centers so that the thought of breaking the social code will prompt unhappy feelings and doing the right thing will make it feel happy and proud.

Disobeying the orders of conscience prompts us to imagine the condemnation of people around us, while obeying them leads us to pat ourselves on the back. The strength of these pangs varies depending on the level of arousal. With conscience, the guilt may range from a slightly uncomfortable twinge to crippling remorse, depending on how bad the deed and how demanding the culture. In the same way, the intensity of all our higher emotions may range from the weak to overwhelming. Usually it is weak because we simply rouse the net of ideas that go with the emotion and only gently pluck on the strings of the emotion centers. We feel patriotism or aardvark-disgust in a low-key, intellectual sort of way. But when we are under pressure, feeling generally excited or sharing our emotions with a like-minded group of people, the emotion can fill us with powerful feelings. As the adrenaline experiments showed, arousal can add powerful spice to even our most cerebral emotions.

So, unlikely though it may sound, people have to learn how to feel higher emotions. Children are taught to pair a framework of social ideas with the right raw emotions. These higher emotions have a quite specific purpose in being carefully tailored by cultural evolution to benefit society. Once planted in our heads, they steer us toward the sort of behavior that helps society flourish. Aardvark-disgust was used as an example to show that any emotion could be created, but since there is no obvious social reason for such a feeling we would not expect it to crop up naturally. On the other hand, disgust with dirt and moral degeneracy clearly have a social value, so it is hardly surprising that we can feel these emotions quite strongly.

Just as with memory control, people do not usually see higher emotions as something they learn but as something they are born with. When parents gently chide their child for throwing the dinner in a guest's face or hitting another child with a plastic toy, they do not feel that they are creating artificial links between correct social behavior and inner feelings so much as coaxing out the finer impulses from deep within their child's mind as seem to be deeply buried at the core of their own being. Likewise, when as adults we manage to stand up to our selfish impulses, we would not feel quite so worthy if we realized that our more charitable instincts had been programmed into us to suit the needs of society, nor would we be as pleased to realize that the warm glow that goes with our little acts of self-sacrifice stems from the pats of approval we received from our mothers for being good. The way children are taught to feel higher emotions is obvious only at the start of the process. As with language, once the child gets the hang of the idea, it will pick up new words and new emotions with ease.

An emotion like guilt may be hammered into children from an early age because it is so important to the workings of society. A society usually has a fairly clear list of dos and don'ts that can be taught as soon as the child learns to speak, but many other finer feelings are learned in a much more subtle way. These tend to be what we might call the spiritual feelings, such as romantic love, pride, artistic feeling, honor, devotion, and fair play. The list also includes more negative feelings like snobbery, cynicism, and distaste. Such higher feelings are not so much taught as absorbed from hearing others talk about them, reading about them, and picking up ideas about them from a hundred different sources. Then we rehearse and practice these feelings—usually during adolescence—until finally they become fully formed.

For example, love is a basic emotion that starts early in life. As children, we are capable of feeling affection, friendship, and a deep attachment to parents, pets, or favorite toys. If we examine these feelings more closely, affection and friendship are the natural pleasure and comfort we find in other people's company, hardly a surprising feeling in such a sociable animal as man. When a child mixes with other children, it experiences a host of rewarding feelings such as amusement, interest, excitement, and fun. Children get similar rewarding feelings from parents and favorite toys, which are compounded by the further pleasure of feeling safe and secure with familiar objects. On the flip side of the coin, children experience the fear and alarm that goes with the absence of these things. The beginning of love in children is thus not a single perfect feeling but a range of quite natural feelings and actions, from happy play in a sandbox to intense distress at losing a favorite teddy bear. Adults can see a common thread to these feelings and give them the general label of love.

As children get older, the sex hormones start to flow and the new element of lust is thrown in with the established feelings, making for a complex brew. To a teenager, the general term "love" now covers the often troubled attachment to parents, the deep pleasure of being with friends, and the lust for attractive members of the opposite sex. On top of this foundation of personal experiences under the general heading of love, we see the subtle influence of culture. Teenagers spend much of their time with romantic books, soulful songs, and slushy movies, which fill their heads with new images of love. Such cultural material is usually treated as low-brow because of its exaggerated picture of all-consuming, perfect, and usually tragic passion. Little is said about real-life imperfections such as headaches, boredom, or bad breath. However, this very exaggeration helps to stretch the ideas of love in the teenager's mind and so give valuable mental exercise for molding a person's own feelings. The typical dress rehearsal stages of puppy love, pop-star crushes, and tragic fantasies should eventually give way to what we call adult love, when a person has blended the different facets of lust, friendship, and attachment into a love that leads to a stable relationship.

If it is hard to believe that modern romantic love is learned, we have only to compare relationships in different cultures.

Modern love is thought ideally to involve the free choice of two equal individuals, but older cultures often have a system of arranged marriages where couples expect to grow to love each other and in strictly male-controlled societies, ideas of love often center on the respectful devotion of a wife living at home.

Love, then, is an emotion that has to be learned. It has natural roots in affectionate social behavior and sexual feelings. Wrapped around these is a cloak of culturally evolved ideas and behavior that vary according to the societies in which people are brought up. Love is not a single pure emotion but a complex pattern of thoughts and feelings that have been shaped to suit a clear cultural purpose. However, having learned to feel love, we can go a step farther and distill in our minds something that feels like the pure essence of love—the sort of sweet feeling that flavors our awareness when we sit on a beach watching a pink sunset or gaze deeply into the eyes of a loved one. This essence is like feeling all the different pangs of love at once without focusing on any specific memories. By distilling all our aches and longings, we fill ourselves with the radiant glow of a curious bittersweet yearning.

This spiritual glow is simply the gentle rousing of the blurred mass of what we might call our love knowledge. Just as we can rouse our net of picnic knowledge to create an unfocused sort of picnicky feeling, so we can stir our net on love into a quiet buzz without dredging any particular memories to the surface. This glow is naturally linked to all our good feelings and so will bring the pleasure center to life.

The ability to summon up the essence of love is still a learned ability, requiring that we first form a rich net of love knowledge, which we then need good concentration and a relaxed mind to stir into life. But once we have learned the trick we can use it creatively as a poet or artist does. We can let the feeling color our minds while contemplating the beauty of crashing surf, a painting, or a geometric theorem.

When we do this, we are perhaps beginning to break away from the strict embrace of cultural evolution. Society shaped up love because it helped to make us self-sacrificing. This love was meant to be targeted at other humans rather than nature, works of art, or fine scientific ideas. However, once we had love for fellow humans, we could learn to associate such "natural" feelings of affection with anything else we wanted. All it took was for groups of people to get together and agree to like the same sort of thing. Today, modern culture supports affectionate feelings for anything from abstract art to submachine guns. Viewed in evolutionary terms—where every human behavior should serve some useful purpose—it is hard to see how some of these modern-day love affairs can be supported. Perhaps these feelings are freaks thrown up by the luxury of civilization and will be ironed out by the pressures of evolution in the long run, or perhaps we are beginning to break free from the control of cultural evolution just as we have already escaped the grip of biological evolution. Whatever the answer, our daily lives are still very much ruled by feelings with clear evolutionary roots.

Love is only the most obvious of these. There are plenty of other culturally evolved feelings that clearly benefit society: devotion, honor, fair play, charity, and all the other noble emotions that push us to be our most selfless. Yet cultural evolution also has its flip side. It also gives us apparently negative emotions such as snobbery and aggression.

At first glance, it seems wrong for negative feelings to be encouraged by evolution but human societies were originally based on small groups which often had to compete with each other to survive, so while we may naturally feel love and loyalty for our own family or tribe, we will also naturally feel fear and distrust for people from rival groups. Cultural evolution sharpens up these negative attitudes in the same way as it does our positive feelings. This leads us to develop "sophisticated" emotions like contempt for people who have failed to get into our group and hatred for people who believe the values of their rival group are better. Modern man has taken these negative feelings to an extreme because he has learned to identify with whole nations and races rather than just small wandering bands of thirty or forty people. By magnifying the numbers involved in such violent emotions as patriotism and racism, he has also magnified the damage done by negative feelings. Nevertheless, the roots of such feelings have a solid evolutionary foundation.

This look at higher emotion shows how big a part society plays in the feelings that flit through our conscious minds. To use a computer analogy, while we are the ones that run the programs we experience as higher emotions, it was cultural evolution that wrote the programs. It should also be clear that it is wrong to try to split the mind into two halves—the rational and the irrational. Traditionally, psychologists have divided the mind into a rational part—which is said to have such thought processes as consciousness, logic, memory, and language—and an irrational part—which is said to give us our unconscious, imaginative, intuitive, and emotional side. But as we have seen, the real split is between the raw and naked abilities of the animal brain and the artificially or culturally extended abilities of the human mind.

Such a split is not a clean separation because every artificial facet has its roots buried in the foundations of the natural mind. Imagination and memory are both the result of language-driven stimulation of the brain's processing surfaces. Self-consciousness is a language-driven extension built on top of the natural foundations of awareness. At best we can say that there is a range from the raw animal to the highly cultural. For example, emotions can range from pure animal fear or rage to extremely artificial feelings such as a love of cubist paintings or "aardvark-disgust." Likewise, logic can range from the natural reasoning power of all smart animals to the driest of mathematical calculations.

The story of man's rise, then, is not that he has steadily become more rational but rather that he has become increasingly a creation of his culture. As civilizations have flourished over the past three thousand years, man has become richer in his supposedly irrational emotions as well as in his intellectual powers. Now that we have seen how cultural evolution created speech, developed the habit of memory, and shaped up the higher emotions, we should turn to self-consciousness itself.

EIGHT

Watching the Watcher

After tracing the evolution of the mind from jellyfish to man, we see that the explanation behind the self-conscious human mind is perhaps surprisingly simple. Evolution equipped us with the mental tools to understand the outside world and we turned these tools inward to observe ourselves.

The ability to look inward depended upon speech. Language allows us to step back in time, if not in space, to gain a vantage point from which to view our minds at work. This makes it possible to treat our own thoughts and feelings as objects to be thought about, alongside objects from the outside world such as picnics and car crashes. With convenient labels such as "mind" or "myself" to help organize our ideas, we can build up a rich knowledge about our own thoughts, actions, feelings, and attitudes.

There still seems, however, to be something more to self-awareness than replaying old thoughts—a shadowy "self" lurking in the background of the mind that remains even after we have stripped away all our thoughts and feelings. Before looking at what gives us this shadowy sense of identity, we need to look at plain unadorned consciousness and clear up the distinction between "raw" awareness and the sophisticated mechanisms of self-consciousness.

The difference between the two has been rather skirted around because most people are used to talking about self-consciousness and consciousness in the same breath. However, as we have seen, self-awareness is a tricky bit of mental footwork—learning to turn awareness around on itself and then living with the net of memories created by this feat. Self-awareness is another recent extension tacked on to the raw animal mind and we need to separate modern tricks from the fundamental ability that underlies them—the naked feeling of being alert and alive that is called awareness or pure consciousness.

We have already talked about such unadorned awareness using labels like the conscious plane or working memory, terms picked to stress the fact that awareness is something that the brain has to do, rather than a passively felt experience imposed on an inert lump of matter. We have looked at how the brain handles nets of perception, memory, imagination, and speech. When we see a rabbit, for example, our retina captures its color and shape and sends coded messages down the optic nerve to the brain. This burst of nerve firings is then splashed across the wrinkled gray folds of the visual lobe and a coin-sized patch of cells at the center of the area lights up in a crude representation of the image. If we lifted the back of the skull, we would not see dancing lights like a flickering TV screen but we would be able to measure a dancing pattern of nerve activity that matched the original rabbit outline hitting the eyeballs.

This first brain image is not the full experience of seeing

a rabbit hopping across a field. The initial patch of activity
is pulled apart by successive rings of processing to extract
details such as color and three-dimensional shape. One part
of the brain may register the whiteness of the rabbit and another part the curve of its stomach. This dissected picture
sparks trails that eventually rouse stored memories of other
rabbit experiences and triggers the vocal sound "rabbit" with
which we tag these memories. All this activity taken together is what results in our conscious experience of seeing
the rabbit.

In the same way, all experience is the result of the brain
dancing to patterns of nerve firing. Dozens of these electro-
chemical storms scud across the cortex every second to create our brightly lighted plane full of conscious impressions
of life. In humans, one of those storms is a net of ideas labeled "me" or "I," which is no ordinary net since it contains
the grand idea that it is the natural target for all the other
fleeting nets in the conscious plane. It includes among its
many stored memories the rather mistaken belief that it is
the observer doing all the watching inside our heads. However, it is the brain that maps out and "consciously" displays each of the nets—including the self-awareness one. What
the self-awareness net actually does is store the habits of
thought that can be used to control the replay of experiences
inside the brain. It may be extraordinary in having this control but it is a very ordinary net in the way it goes about the
job. Everything going on in the brain is in fact the result of
the well-organized firing of nerve cells, and self-awareness is
not a special type of firing but simply a new sort of firing
pattern.

Our minds are the end result of the twittering of billions
of nerve cells within the handful of pinkish mush that sits
inside our skull. Although we may accept this fact intellec-
tually, it is difficult to compare the mechanical explanation

to the living experience. It is hard to do real justice to our intense experience of life and capture with words the colors, shapes, surges of emotion, and buzzing ideas that flood our conscious plane.

For our body's other organs we have everyday analogies that give us a good mental picture of how they might work. When modern man discovered that the heart acts like a pump to push blood around the tubes of the body, people quickly got used to the idea since they knew about pumps from everyday experience. Most of the body's other organs turned out to be equally easy to understand through the use of ordinary metaphors; For instance, we talk about the kidneys being filters for straining the blood, the eye being like a camera, and the lungs like sponges that soak up air. But there are no such straightforward analogies for how the brain works. In this book, we have used the metaphor of a TV screen in which the picture is made up of thousands of flashing spots—with the same spot being part of, say, Clark Gable's left ear one second and a red-tile roof the next—but a TV screen needs a viewer to watch the fleeting images that pass across it. We have no experience of a TV screen that knows how to watch itself, so we find it an impossible idea to picture clearly.

The computer might be a closer analogy than the TV screen for the way the brain consciously experiences life. After all, computers have the equivalent of biological circuitry in their intricate wiring and they also have the etched patterns of memory on their magnetic disks, which store both a computer's experiences—its data—and learned habits of thought—its programs. They even have something like self-awareness in the way that they have special systems software to switch around between programs and notice problems as they crop up. But today's computers still fall a long way short of the complexity of the brain. A computer has no match for the sophistication of society's programming methods or the sort

of hardware in the brain that can handle the flood of information flowing in from the senses.

Taking a closer look at our raw consciousness, we find no
sharp boundary cutting off awareness from its foundations.
We have seen how the flood of sensation arriving at our sense
organs is channeled and filtered to form the rising froth of
conscious experience. Normally we let this tide rise as high
as it will go and live our conscious lives at quite a rarefied
level of perception. When we look at a mug of coffee, for
example, we get the feeling of seeing the whole thing rather
than a mass of details. The cup's handle, its contents, its
rim, the steam rising off it, the pattern on its side, all are
fused together into a glowing impression in our minds. We
are aware of the detail but it is only a vague awareness compared to the sharp sense we have of seeing a whole mug.

To break this glowing mental net back down into its component parts, we need the help of language so that we can
consciously zoom in on details. With this artificial aid, we
can zero in on the cup's handle, then force ourselves to keep
on going and look more closely at the curve of the handle's
surface, the reflections of light playing on it, and so on, until
we reach the limits of what our eyes can pick out. Indeed, if
we know what we are looking for, we can get down to the
black and white lines of contrast that run along the edge of
the handle when we hold it up against a bright background—
the same artificial edge we saw around skyscrapers and books.
We can travel right back down the pyramid of processing that
forged the total image of the mug and catch out the little
tricks of processing as they take place out at the retina.

To do this, we have to fight against the natural tendencies
of the brain. Evolution has shaped the brain to boil down the
buzzing confusion of life to its simplest possible shapes and
patterns, so our sensory pathways always fight to push our
awareness of objects up to the highest level of organization

possible. We will always tend to be aware of the whole
steaming mug of coffee rather than a jumble of edges, reflections, and curves. Likewise, we will normally see a person's
whole face rather than a jumble of lips, nostrils, and eye-
brows. Everything in life—trees, boats, cars, anything—will
live in our minds as complete objects rather than as assem-
blages of parts because that is what the sensory pathways are
designed to achieve.

Of course, part of an object can draw attention to itself. If
an eyebrow is raised or a tree limb bends, we are drawn into
focusing on it but we still see this feature as a whole. We do
not suddenly see the eyebrow as a collection of hairs or the
limb as a collection of leaves and twigs. Our sensory pathways are still trying to push our bright awareness of an object up to its highest sensible point of organization.

Thinking about awareness in this way perhaps gives a
clearer idea of what it is to be conscious. The flood of infor-

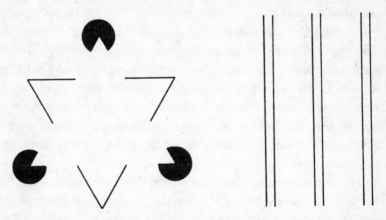

*Because our visual processing struggles to present us with complete
shapes, we see a triangle that isn't actually there.*

*Likewise, when looking at this collection of lines, our brain groups
them into three bundles rather than seeing them as a scattered
assortment of six lines.*

mation pouring in through our eyes, ears, and other senses is
boiled down into organized patterns of nerve firing which scud
across the surface of the cortex. Real-life objects, such as coffee cups, are mapped onto the brain as a web of cell activity.
Our consciousness is not then the result of some inner observation mechanism stepping back to inspect ghostly mental images of a cup. This cup-shaped web of nerve firings is
itself a froth of understanding whipped into life on the brain
surface.

This may not sound like a convincing explanation of consciousness because it is hard to escape the feeling that what
lives inside our skulls is too bright, too sharp, and too effortless not to be the result of something more than a pattern of
connections between gray nerve cells. But, as with our power
of imagination, it is easy to overlook the flaws in what we
experience. When we picture a green apple in our minds, we
believe it to be a rich and accurate image. It is only when
we pick up a real apple that we realize the imagined experience was a pale imitation. But should we then think that the
real apple gives us the most intense and complete experience
we could have? Perhaps if we had sharper eyes and a bigger
area of brain to map the image on, the conscious experience
would seem more real still. The only reason for feeling that
our present level of awareness is the ultimate is that we have
not got the mental equipment to create an even richer experience.

The same principle applies in our dreams. When we sleep,
we feel we have completely lifelike dreams, but if we examine our dreams more closely, we see that—like remembered
meetings in sunlit corridors—the experience is flawed and
disjointed. We telescope what seems to be hours of action
into a jumble of fragmentary images but because we are asleep
and shut off from the usual flood of images from the senses,
this dream world seems to us to be the most vivid possible.

In fact, the nets that make up our dreams probably employ no more nerve cells than we use in our waking imagination, yet with no rude intrusions from real life to put our pale and fragmented dreams into perspective, we are left believing them to be intense experiences. The brain is designed to grab what input it can and then boil it up into a froth of understanding. Because it uses everything at hand, and nothing is left out, the results are accepted as the most complete experience of life possible.

Consciousness is thus to be found in the patterns of firing that skate across the surface of the brain. Some of these patterns always seem to be in sharper focus than the rest so while we may be vaguely aware of many things at any one moment, there will always be one or two things that stand out, catching our full attention.

Attention is a feature of the mind that is often overlooked—probably because it is so fundamental that it hardly seems to need explaining—but the way that certain impressions manage to claim center-stage in our mind is the key to understanding how awareness works. Reading this book in a comfortable chair, your mind will be made up of the ebb and flow of many nets: the sight of the page and also the room as you occasionally glance around; the distracting sounds of your surroundings; all the various itches, twinges, and pangs of an ever-complaining body; and finally a series of inner thoughts and images—even of yourself sitting in a chair while reading. At any one moment, only one of these sensory nets will be caught in the sharp central spotlight of attention and the rest will lie in the dim background, awaiting their turn.

This sharp focus is extremely restless, usually staying in any one place a matter of split seconds. Even an apparently continuous act, like reading this book, is actually a stream of small interruptions as our attention is distracted and we have to drag our mind back to the page. We may stop briefly to notice a slamming door or to scratch an itchy leg before again turning our eyes back to the book.

It may seem a handicap to have such a restless and narrow sort of awareness. Might it not be better if evolution had equipped us with minds that were equally alive to everything within sensory reach rather than having to illuminate the world with a flitting pinpoint of light? However, pinpoint attention makes evolutionary sense. The body is a bag of muscles and organs that all have their own needs but must be made to work in a coordinated fashion. After all, the need for coordinated action was the reason why animals evolved a brain in the first place. Animals wanted a nerve center that could take account of the situation both inside and outside their bodies and then plot the best course of action. While the brain does this job for the body, it is attention that does this sharp focusing job within the brain. It chokes the general blur of sensory and bodily information down to a pinpoint focus so that an animal can attempt to do only one thing at a time. Because a rabbit will focus most clearly on either the stalking fox or the long lush grass—but not both at the same time—it will be able to decide what to do. If it tried both to eat and run or to do a little of both, the results would obviously be disastrous. Because attention is also restless, as well as pinpoint sharp, there is less danger of an animal concentrating too much on one thing. Evolution made sure that the rabbit would be twitchy enough to nibble only a few blades of grass before checking nervously on its surroundings.

The way attention has so far been described makes it sound like a roving inner spotlight that we turn toward the firing nets we are most interested in. Once we light on a net, our attention triggers it brightly into action and so switches up its volume to full blast. There is something in this picture of attention, but it really applies only to the language-driven afterthought tacked on top of raw natural attention—what we call concentration. It is a misleading way to think about raw attention because there is no inner hand to direct the shifting spotlight, any more than there is an inner eye viewing the patterns splashing across the visual areas.

The active process we call attending is the final result of the natural self-organization that goes on within the brain's processing surfaces. What we think of as the sharp focus of attention is one net flaring up brightly among the glowing background jumble of its fellow nets. Like a match tossed on a heap of smoldering coal, the net bursts into life and forms the most intensely felt pool of consciousness within our heads at that moment. A particular net is selected for this bright central focus as the result of a complex committee decision. First of all, the sensory pathways have a major say in the matter and at any one instant all the body's sense organs will be clamoring to get their messages through to the brain, with each sense trying to dominate our attention by burning most brightly in the conscious plane.

When the sensations are finally splashed across the wrinkled cortex, each net rises up a pyramid of processing until it strikes an answering echo of recognition from matching memories. The noisier an arriving net is, the more it appears to grab our attention and the more urgent seem our attempts to recognize and judge the intruder. Our attention feels sharpest at exactly the point where we make the match between an arriving net and our warehouse of knowledge, when we have just recognized what it is that we are sensing. We feel this clarity, for example, when we have just realized that a funny feeling is an itch on our leg that needs scratching. Or when we have just understood that a sudden noise is the clanging of a bell—which is, anyway, probably a false alarm that we can ignore. Both the itching and the ringing start out as vague stirrings on the fringe of awareness. The process of

recognition then follows swiftly to create the sensation of
sharp understanding.

What makes our focus of attention feel so pinpoint sharp
is that our brains seem to be able to handle only one act of
recognition at a time. We do not recognize everything in a
room at once. We first recognize the global net of the "room";
then we cast our eyes around and start picking out little things.
Attention is a serial or step by step process. We rapidly flick
our gaze around our environment, pausing for a split second
to recognize and understand each net that is thrust into view.
It is rather as if all the clamoring nets reach a narrow bridge
that leads to the memory zone. Only one net at a time can
cross halfway to shake hands with its matching store of
memories from the other side of the divide. The nets all have
to take their turn to be recognized and properly linked up
with their matching partners in memory; then they are quickly
jostled off the bridge by the horde of other nets waiting their
turn. But once the connection has been made, we feel we
have fully understood that particular part of the world. This
understanding lingers on in working memory before it eventually fades into the gray background of the mind.

Until we have put something under the spotlight of attention and forged links with memory, we do not feel that we
have had a sharp conscious experience of that object. Anything that we have not got around to focusing on stays a
vague blur on the fringes of our awareness, yet while these
fringes appear murky to us, they are definitely not unintelligent. Our brain is quite capable of handling even very difficult tasks without needing the limelight of sharp consciousness.

For example, we can drive a car while at the same time
carrying on a conversation with a passenger sitting alongside
us. Our mind can be fully taken up with the job of speaking
and listening. We will be driving on autopilot, only vaguely

aware of doing such things as changing gears, avoiding pot-
holes, and keeping an eye out for traffic police. This apparently dangerous lack of attention rarely gets us into trouble
for our brains are so well trained to drive that we do not have
to attend to every detail. Like hitting tennis strokes, the
driving skills have been programmed into the movement
center of our brain, the cerebellum, and we can rely on our
brain to do most of the driving automatically. The cerebellum bulge still needs to feed off the consciously experienced
image of the world that is being mapped out higher up in the
brain, in the cortex, but it does not need the roaming focus
of attention to bring sharp consciousness and understanding
to the scene. A general background awareness will do.

On the other hand, when we speak to our front-seat passenger, the conversation is always new, so we have to think
clearly about what we are saying. We need to use the sharp
focus of consciousness to check that we are making sense as
we speak. While we are driving along on autopilot, we rely
on the natural attention-alerting powers of our senses to drag
us back to concentrating on the job of driving if something
dangerous crops up. A dog dashing out in front or a difficult
corner will bring our full attention back to the road. Our
chatter will be cut off in mid-sentence. Sharp attention is
reserved for the tasks that need the most help from our vast
warehouse of memory. The brain is intelligent enough to deal
simultaneously with a lot of other routine or well-learned
actions, even though they are not illuminated by the sharp
spotlight of attention.

We have seen how attention describes our most intensely
felt experiences. Awareness is strongest when brightly burning sensations strike a chord of understanding within our
memory banks. Putting it another way, our conscious experience of life is nothing more than little flashes of recognition going off in our heads all day long. It is because our

brains are filled with a nonstop succession of matching between sensations and memories, like beads on a string, that
we get a feeling of being continually conscious. As soon as
the brain stops this pairing—as when we go to sleep—we lose
the feeling of being aware.

Again, this underlines the fact that our sense of being conscious is an active process rather than some phantom object.
Consciousness is the label for what we experience when the
brain repeatedly matches incoming nets and memory to put
together a string of understanding. As soon as the brain stops
working—or even drops below a certain threshold of activity—consciousness evaporates. Consciousness is not something inside our heads that does all our experiencing of life
for us. It is simply the description for the stream of recognition flashes that take place as long as our brains are at work,
the string of mental "ahas" of recognition matches being made.

Returning to the process of attention, we have seen how
the apparently roving spotlight is really the bright flaring of
nets as they connect with memories. There is no inner mental searchlight—that is only the easiest way to describe what
we experience. A better picture is the jostling queue of nets
trying to shove their way into the center of consciousness so
that they can cross the bridge to memory and understanding.
But even this image implies that nets have to shuffle around
to a particular part of the brain, when what happens is that
various nets are splashed across different parts of the cortex
and the brain is dominated by a particular net because its
firing becomes raised enough to start triggering memories.

It is not clear how the brain raises the firing levels, but we
can remember, from the way the eye works, how nerve cells
in the retina can turn the firing of neighboring cells either
up or down to exaggerate the light and dark contrast of a
sharp edge. The same thing could happen on the brain surface where a noisy new net has the energy to switch off qui-

eter neighboring nets. By damping down surrounding activity, the new net will in effect draw attention to itself by its contrast with the gray background and will enjoy a brief few seconds in the limelight until a fresh net arrives with enough energy to switch it off. When this happens, we will experience what we call a shift in attention; that is, a brightly burning net in one corner of the brain will be overshadowed by the arrival of a fresher net in another corner. Every second, another salvo of nets will land, causing fresh explosions of understanding to ripple across the cortex.

This describes natural attention, but as usual, humans have built a language-driven extension on top of this raw ability. We use language to power the artificial ability called concentration, which is the learned skill of being able to force ourselves to keep returning to a task despite a natural tendency to get bored and distracted. Indeed, distractions should really be defined as the things that we find naturally interesting in life.

The way concentration works is that language allows us both to set ourselves long-term goals and to remember to keep returning to them. For instance, when we watch a fly crawl up the window pane while at work, we come around with a start and tell ourselves to get back to the job in hand. Our inner voice can also be used to prime ourselves to look out for things we want to find. When baking a cake, for example, we see that we need flour and sugar and with these words in the forefront of our minds the packets will leap out to grab our attention as we scan the kitchen shelves. Words form the links that lead our attention back to where we want it focused, whether we are reminding ourselves to return to a long-drawn-out task or keeping our eyes peeled for something in particular.

Intense concentration is very tiring because we are fighting against the natural pull of distractions. This was soon discovered by the Skylab astronauts whose every minute had been timetabled for doing something useful because of the costs of being in space. But despite their motivation, these astronauts found that their minds kept wandering off if they did not rest regularly.

Concentration may be hard work but the skill was a big advance for humans. Stone-age man would have needed patient attention to make tools and to plan hunts. Even more important, an ability to control our attention would have been vital for self-awareness. Mental events are ghostly, slippery things. It would have been hard for early man to focus on the workings of his own mind for long, given all the distractions of the busy world outside. Good concentration would have been needed to contemplate our thoughts for even a few seconds, given the tide of sensation that floods into our brains, demanding to be noticed, every moment of our waking days.

If self-attention was a key step toward self-consciousness in man, yet was a difficult task because the outside world was always more exciting, man must have taken it because looking inward proved to be a valuable habit and so was encouraged by the guiding hand of cultural evolution. Either there must have been a strong social reason for making man concentrate on his own mind, or else inner concentration was a lucky side effect of other things that society was making early man do. Before looking at the social reasons for having self-awareness, we should first consider the limits to our self-knowledge, to see exactly how self-aware it is possible to us to be.

The main limit of self-awareness is an obvious one. Everything we can be aware of, we should also be able to be self-aware of. If self-awareness is stepping back to look at our own mental processes, then every net that flickers across the conscious plane is a candidate for being replayed. By the same token, what lies outside awareness cannot be viewed directly by self-awareness; for example, we are not conscious of the mechanics of sentence forming, the workings of recognition, or the chemistry behind swings of mood. We are only aware of the effects of these processes as they spark conscious—or sensation-based—nets in the mind.

This is not the same as the classical Freudian "unconscious," which was treated like a part of the brain walled off from the conscious self. It was thought to be a mental cesspool full of repressed sexual desires and death wishes, which our conscious self would struggle to keep in check but which occasionally would erupt and burst through into consciousness, showing up as dreams, slips of the tongue, and neurotic behavior. Freud's belief was that one or two traumatic childhood incidents caused later mental problems, as if the nasty memories get locked away in the subconscious but later disrupt our lives with their kicking and screaming as they try to break out. From what we now know of the molding of our minds, we would not, however, expect our personalities to be shaped by one or two childhood incidents. Instead, thousands of small childhood events would gradually etch the memory surface that in later life drives our behavior along in characteristic patterns of thought. These thousands of small memories would be unconscious not because they are locked out of consciousness but because they are too faint and numerous to single out easily. The memories texturing our mind might include, for example, a general impression of whether our parents were relaxed or anxious when we made our first stumbling steps as toddlers. But while each individual memory might be faint, together there would be enough of them to combine to form a richly patterned mental surface. This underlying texture would later have a powerful effect on the fleeting nets of sensation and thought—perhaps tending to make us neurotic or disturbed.

When psychoanalysts use techniques like hypnosis and

therapy to shake our memories hard, what comes out are a few vivid occurrences, which tend to be seized on as the explanation for all our problems. Yet while a traumatic incident might deeply scar the memory surface, its influence over our lives would still probably not compare with that of the thousands of smaller memories lying unrecognized in our minds. And more important for how we view the unconscious, we would not actively repress any of these formative memories. It would simply be too difficult to prize them out from their blurred surroundings. The true unconscious consists of the nonconscious parts of the brain—the various brain pathways not involved in the glowing map of the outside world upon which consciousness is based.

The emotions are a good example. When erotic or threatening sensations flicker across the conscious plane, they spark messages to the special emotion center at the heart of the brain known as the limbic system. Once stimulated, the limbic system can stoke up the body's fires to ready it for energetic action. But although this metabolism raising takes considerable coordination, we are not directly aware of the activity of the emotion center as it goes about its job. The mechanics of the process are as much out of sight as the mechanics of forming the sentences we speak. We can only observe the results of the limbic system's activity after it has triggered the body into action, when through consciously experienced sensations, like a pumping heart or sweaty hand, we can guess that something important is going on. Self-awareness thus has an absolute limit. It cannot directly examine anything that is outside the thoughts and sensations mapped onto the conscious cortex. The most it can do is detective work to deduce what is happening elsewhere among the brain's many control centers.

There are other limits to self-awareness which are not quite so clearly defined. As already seen, an example of these blurred limits to awareness is the difficulty in fighting against the rising tide of perception. Because the sensory pathways are shaped to make our impressions of life as organized as they can, it is a struggle to get back down to the fine detail at the base of the pyramid. Our self-awareness of events in the conscious plane is limited by our skill in breaking nets back down into their detailed parts.

A further type of limit is the brightness of consciousness. When we are tired, it is as if the voltage in our brains has been turned down. Our awareness of life dims generally. On the other hand, when we are excited, the voltage is turned up and our thoughts dance with energy; our senses are razor-sharp. To be at our most self-aware, we also have to be at the peak of our general awareness of the world around us.

Given the physical limits of awareness, we can now look at what cultural evolution has made of the possibilities that exist for self-knowledge. Cultural evolution, being what it is, would have encouraged the habit of self-awareness only so long as it served a useful purpose in the social life of humans. It would not necessarily exploit the full potential of the human brain for being self-aware. The social mold forged by cultural evolution would teach humans to go only as far as suited its purposes, which makes it important to ask what social good it does to be self-aware at all.

Being self-aware has both good and bad effects as far as society is concerned. The bad side is that, taken to extremes, the habit of looking inward could be time-wasting and lead to an unhealthy self-obsession. From cultural evolution's point of view, a race of introspective and selfish individuals would be a disaster. Society's logical aim would be to turn out people who slot smoothly and efficiently into groups and any self-awareness skills of the group members would be to help each individual check that they were doing their best to meet the group's goals.

From this point of view, our ability consciously to put on the correct social mask to go with the social occasion is an important reason for self-awareness. With the increasing complexity and specialization of modern culture, this need has probably led to a sharper sense of self than ever before. We are so aware of having to keep up a mask in public that our own inner world of private thoughts is thrown into sharp relief. Just as a tight pair of shoes can make us uncomfortably aware of our feet, the way that society forces us to wear a mask seems to have inadvertently made us more aware of the inner self we are masking.

Another reason why society might have made us self-conscious is that society needs us to be self-aware so that we can police our minds for antisocial and disruptive impulses. Society is built out of a delicate tissue of relationships and actions that can be ripped to shreds by the simplest impulsive behavior, such as giving in to the urge to hit someone in the face for annoying us or telling our boss at work just what we think about him. Our selfish instincts throw up all sorts of thoughts about what we want to do or say, but self-awareness keeps an eye on the thoughts flickering through the conscious plane and when it recognizes an antisocial urge welling up, it can quickly step in to nip the impulse in the bud. We saw this happening with conscience but this self-conscious policing of behavior does not have to be as blatantly emotional as conscience. Our day is spent stifling the smallest bad-mannered impulse or reminding ourselves to get on with the tasks expected of us by our jobs or position in society.

Apart from this moral-policeman role for self-awareness, early man's other important reasons for focusing on events in the conscious plane would have been to use tools and to make plans. Early man would have needed to focus very clearly on the actions of his hands and body to chip away at a stone ax or to throw a wooden spear. Such skillful movements need close monitoring. Each blow against a flint has to be carefully planned and precisely struck, while the throwing of a spear needs a careful aim and patience to wait for the right moment to let fly. Skilled action needs the same sort of self-conscious review as skilled social behavior. Planning ahead or foresight was another valuable skill that would have demanded self-awareness in early man. We need to focus inward on our thoughts to remember our past and create imaginary pictures about what we should do in the future. Early man would have had to evolve these mental skills before he could organize hunting parties or trips to find the right flint for toolmaking.

There were, therefore, several reasons why cultural evolution should foster certain types of self-awareness, and once the self-awareness ball started rolling, its momentum probably took it farther than really suited society. Society obviously benefited from early man's ability to turn his attention inward to monitor his own thoughts, because it got him making tools, thinking ahead, and curbing his antisocial impulses, but once man had got into the habit of watching himself, it was inevitable that he would build up memories that led to a rather troublesome sense of the self. Self-attention could not help but cause a blurred knot of self-knowledge to accumulate, which would then automatically power a sense of self-identity.

So far we have been talking mainly about self-awareness as the process of looking inward to observe and guide thoughts, but as mentioned earlier, there is also a shadowy feeling of self that appears to form the very core of our being. Remembering how nets of memory behave, however, there is nothing mysterious about this sense of self. We have seen how dense knots of memory can be formed by the blurring of many similar or related memories—for example, our net of knowledge about picnics is made up of many experiences, facts, and images and it can be used by us in many ways. We can dig out memories to re-create one particular childhood picnic, use the blur of memories to create a completely imaginary picnic scene, or use the net as a sounding board to answer such obscure questions as whether we are more likely to find jam or sardine sandwiches at a picnic—or, of course, we can simply use the congealed net of picnic memories to recognize what a group of people are doing in a field.

Similarly in self-awareness, the simple habit of looking inward automatically creates a blurred net of memories about all the thoughts we have observed. Every time we analyze our reasons for losing our temper or notice the social front we put on at a party, we are weaving another strand into the general net of ideas we carry about our own mental lives. This net will then behave like any other net of memories. We can dig into it to isolate specific memories or answer factual questions; for example, it will provide the detail that allows us to explain our behavior or act consistently—telling us to stick money in a collection box because we are charitable personalities or not to get too excited at a football game because we are cool customers. More important, we can tug generally on the whole net and arouse a blurred sense of "me-ness." Instead of dredging up any particular memory, we can stir up the vague glow of a knotted mess of memories, which we experience as our sense of identity or self.

The sense of self is never far away from the conscious plane. We occasionally feel so absorbed in a book or film that we forget ourselves and for a few seconds at least we lose sight of the fact of our own separate existence. But usually we go through life with the net about the self always glowing dimly on the fringe of awareness. This is because the natural processes of recognition keep tugging the net of self-knowledge to life. The very act of being aware automatically strikes a chord with this net of memories and drags it back into the conscious plane. All day long we are continually recognizing the fact that we are aware. Our mind keeps rediscovering the fact of its own existence through the simple act of self-recognition.

Self-recognition is such a constant event in our minds that we treat it as a fixed part of consciousness. We believe that our sense of self will always be there and find it hard to imagine that we spend all day rediscovering ourselves. Yet when we wake up in the mornings, for the first few seconds we are simply blurrily aware. Then our patch of self-awareness memories recognizes the familiar activity of awareness and sparks into life. We pull the world into focus and suddenly feel that we have become our true self-aware selves again. Once triggered, the self-awareness net brings with it all the habits of inner control. We can start thinking about what we should do for the rest of the day or dredge up memories of what we did the night before.

Self-awareness, then, is a relatively trivial trick of the brain. Once we are in the habit of watching and guiding thoughts, we automatically start building up a dense knot of self-knowledge, which is continually jerked back into life because we cannot help but keep recognizing the fact that we are aware. Our sense of self is thus kept bubbling away on the fringes of our awareness, all day long.

A further twist to the blurred net that creates our sense of self and makes it different from all the other nets in the brain is that the net of self-knowledge assumes that it is responsible for everything happening around it in the mind. Self-awareness is big-headed. The net believes it is in charge of the whole show and claims the credit for everything, from the flexing of a finger to the forming of a sentence. It claims that pain and happiness are things that it feels, sights and sounds are what it sees and hears, and flashes of inspired

thought, feats of athleticism, and artistic brilliance all flow
from the activity of this fairly small mesh of cells. In fact,
the self-awareness net is mostly a bystander to the events
taking place in the brain. At best, it plays a crude part in
influencing them. Like a clumsy colonel, it can bark out a
few orders that are then turned into action by the sleek ma-
chinery of the brain. It is quite good at vetoing courses of
action suggested by other parts of the brain: A simple yes or
no can cause us to hold back a trembling fist or to go ahead
with a venomous sentence. It can also drag attention back to
a task that needs doing or keep prodding a chain of thought
along to a conclusion. But the net of self-knowledge is not in
sole charge of the mind. It is just a particularly opinionated
patch of memories that knows a few tricks to lead the mind
down certain well-trodden tracks.

We started with an image of self-consciousness as a won-
derful and unfathomable gift of the human mind and we have
ended with a more prosaic picture. A habit of thought caused
us to build up a tangled net of memories about ourselves,
which then started to claim for itself some grandiose powers
within the brain. Every time the cerebellum cracked a win-
ner in a tennis match, the net of self-knowledge would note
the fact and include it in its vast collection of ego-boosting
memories. But this net is a net like any other and is dependent on the brain for its survival, just like any other net.

Creating an inner driver has nevertheless made an enormous difference to the humble hominid brain. It opened the
door to cultural evolution so that we could steadily evolve a
better mind over the past forty thousand years. By learning
to turn our attention inward and focus our awareness on our
own minds, we have come up with all sorts of new tricks,
ranging from personal memory to emotions. Nor is this necessarily the end of the story. With human culture in the
twentieth century developing at an accelerating rate, more

changes to the human mind are possible. Indeed, once we have realized how plastic the human mind is, we can think about taking its shaping out of the hands of cultural evolution. Rather than relying on the blind forces of cultural evolution, we could consciously decide where we want to go with our minds and how we are going to get there. But deciding what we want may prove more difficult than the going and getting.

NINE

Truly Self-conscious Man

Judged by evolution's usual patient time scales, self-aware man has sprung up almost overnight. After hundreds of millions of years of small, creeping advances in intelligence, a lucky branch of the ape family discovered speech and raced away to develop an inner control over the brain. But does the story stop there? What is the future of self-consciousness?

At the moment we seem to have all the self-awareness that we need. Society does not seem to have any special interest in further polishing our ability to be self-aware. The self-consciousness that we have is almost a side effect of the skills that cultural evolution wanted, skills for specific jobs such as putting on social masks or using foresight to plan ahead. Because society only indirectly gave us an ability to be self-aware, there may still be a lot left undone that would greatly improve our powers of self-consciousness. Perhaps all we have to do is understand the true nature of self-consciousness and set about making some of these improvements.

Probably the biggest factor stopping us from being more self-aware is the blinkered idea that our present self-consciousness is all that there can possibly be. If we accept the idea that self-awareness comes naturally or is something mystical, then we will never try to find ways to improve our knowledge about our self. We feel that we are already properly self-conscious. But all this means is that we are aware of the fact that we are aware. We do not understand the detail of how our mind actually works. Neither are we aware of the precise limits of our consciousness or how we actively push thoughts, memories, and imagination through our conscious plane. And we are not aware of how our minds are molded by our societies, so we are not alert to the hold that society has over the way we think.

If we understood all these things, we should become much more self-aware. We would have a proper framework for gathering the ideas about ourselves that go to make up our blurred knot of self-awareness memories. We would know what sort of things to expect to find going on in our heads. We would be looking out for details that we probably would have overlooked before, like word-driven memory searches and the rediscovery of self-awareness when we wake up each morning. And we might also start analyzing some of our thought processes more accurately. After doing this for a while, we would find that we had developed a richer and better structured sense of self-awareness.

This is not to say that we would move on to some higher plane of consciousness where we would live in a blissful mental haze. The conscious plane is tied to the physical world of the brain and the only way it can be made brighter or more intense is through a raising of the brain's metabolic thermostat, such as we experience when excitement makes us keenly alert. Self-awareness is an activity that takes place on top of this conscious plane, so an improvement in self-awareness means better understanding and control over what takes place in this mental arena rather than any change in the physical foundations of raw consciousness.

The difference is more like that between a golfing beginner and a club professional. The beginner believes that he plays golf and can manage a few wild swings, though when the ball flies into the trees, he is not sure why other than that he might have mis-hit it. The expert golfer, on the other hand, plays the game with practiced skill. Every shot involves thought and planning, and if it goes wrong, the professional will probably know that he kinked his elbow on the backswing, misjudged the wind, or whatever. Self-awareness is similar in that everyone believes that they can play the game but few put much practice into learning to do it well.

Most people bother only with the broad outlines of their minds. We notice the general glow of our self-awareness, the social masks we often put on, and the powerful surges of emotion we sometimes feel, but most of us ignore the trivia of our minds. Given how hard it is for us to concentrate on the inner world of the mind while we are faced with the powerful tug of distractions from outside, it is not surprising that we do not routinely watch to see how we form the sentences we speak, how we decide to lift a hand to scratch our head, or how we keep bringing our concentration back to the printed page after a distraction. We have neither the energy nor the interest to force ourselves to concentrate on the minute detail of what goes on in our heads. Perhaps, however, having seen how important it is to build up a good feel for this detail in order to power our sense of self, we may decide it is worth starting to make the effort.

What about the future of self-awareness? We have already looked briefly at the possible effect of computers on extending our powers of thought. The relationship between man and machine could eventually go much deeper than that. The natural trend will be to off-load the mental tasks that our human minds are not very good at onto the hardware that surrounds us. We are already starting to do this: Schoolchildren use calculators instead of learning mental arithmetic; students have microfilm and electronic databases so that they do not need to memorize so many raw facts. The better the methods become for getting stored information back out of these machines, the more we will come to rely on them as an extension to our own memories.

Computer companies are beginning to develop many other aids to the brain. We are starting to see what are called decision support systems, which analyze problems and offer the user a list of sensible solutions. Other systems rigorously check through the logic and model the likely outcome of the user's own suggestions. Although only big businesses will initially be able to afford these systems, the falling prices of hardware mean that all computer advances should eventually become cheap enough for the individual user to buy.

Many more aids to the mind will probably emerge out of computer technology, and as we get used to each new invention, we will be subtly extending the power and range of our minds. At the moment, we have to live with almost naked minds. We have extended the range of our possible experience of life by books, television, and hi-fi equipment, but not many people yet have access to the hardware that improves their ability to assemble facts and ideas—or that helps them work through those facts to reach sensible theories and conclusions. There can be little doubt that such advances are just over the horizon. Just as today's man may feel mentally naked without a daily newspaper and a calculator in his pocket, tomorrow's man may feel exposed without computers whispering facts and suggestions in his ear.

We could say that the story of the human mind so far has been one of constantly improved software running on the same old hardware. We have had a constantly evolving language-driven mind running on the same three-and-a-quarter-pound brain for the past forty thousand years or more. Perhaps within the next decade or so technology will start improving our hardware as well. The question then becomes, what sort of minds are we really aiming at? If we are going to off-load many of our mental tasks onto the hardware surrounding us, what sort of mind do we need at the center? Obviously, there will be a premium on adaptable and flexible minds. In the old days, a set of beliefs and mental habits easily lasted the owner a lifetime, but the pace of modern science and society is becoming so fast that we might need to become experts at remodeling the ways we think. Perhaps we will need computers to act as a surrogate society which can rapidly retrain us in new styles of thinking.

Moreover, since most people still believe that the mind is fixed and unchanging, rather than plastic and molded by society, they have not yet started to think about what sort of improvements they could make. Perhaps we will decide we want to make the effort to become more self-analytical so that we can break free from the hold that society has on the way we think. Maybe we will want to decide for ourselves exactly what sort of minds we should have. Freedom may come to mean not only freedom to do what you want but also freedom to think in a way that is independent of the velvet grip of society.

Breaking free of society's grip might prove traumatic, however. Humans are by nature gregarious. We have seen from man's evolutionary history how he is even more social than his close relatives such as the chimpanzee. Even the amount of self-awareness we now have causes many people much unhappiness. They feel isolated inside their own minds, cut off from the warm embrace of the humanity around them. They would hardly want to do anything that increased this mental isolation. Instead, they might want to use discoveries about the mind and technology to get closer to other people. They might prefer to work on the sort of emotions and social masks that would make a more tight-knit and well-ordered society possible. Having realized the loneliness that real freedom can bring, they might prefer to have real friendships instead.

Most likely, our future will be a mixture of these two trends. Some people will pursue freedom of thought and others the warm embrace of society. Both will have a real choice once they understand how minds can be shaped and improved, for real self-knowledge should bring us genuine self-control. Whatever happens during the next fifty years or so, it should certainly add an interesting new chapter to our story of the mind of a self-conscious ape.

Bibliographical Notes

1. The Balloon-Headed Ape

Page 15: For a good plain description of the statistical nature of evolution: *The Blind Watchmaker* by Richard Dawkins (New York: Norton, 1986).

Page 15: For equally good roundups of ape and hominid evolution: *Lucy: The Beginnings of Humankind* by Donald Johanson and Maitland Edey (New York: Warner Books, 1982); *The Monkey Puzzle* by John Gribbin and Jeremy Cherfas (New York: Pantheon, 1982); and *Origins* by R. E. Leakey and Roger Lewin (New York: Dutton, 1977).

Page 17: Brain energy consumption: A classic study is "The Determination of Cerebral Blood Flow in Man by the Use of Nitrous Oxide in Low Concentrations," S. S. Kety and C. E. Schmidt, *American Journal of Physiology* 143, pp. 53–56 (1945). See also *Energetics and Human Information Processing*, edited by G. R. Hockey, A. Gaillard, and M. Coles (Dordrecht, Netherlands, Martinus Nijhoff Publishers, 1986), and *Nutrition and the Brain* by R. I. and J. J. Wurtman (New York: Raven Press, 1977).

Page 18: Ape numbers: Figures from the World Wildlife Fund.

Page 18: Nimble fingers: "The Seed Eaters: A New Model of Hominid Differentiation Based on a Baboon Analogy," Clifford Jolly, *Man* 5, pp. 5–26 (1970).

Page 19: Bipedalism: "Evolution of Human Walking," Owen Lovejoy, *Scientific American* 259 (5), pp. 82–89 (1988).

Page 20: Australopithecus diet: *Lucy*, Johanson and Edey, p. 363.

Page 21: The ice ages: *Ice Ages: Solving the Mystery* by John Imbrie and Katherine Palmer Imbrie (Hillside, N.J.: Enslow Publishers, 1979).

Page 22: K and r: *Lucy*, Johanson and Edey, pp. 324–344.

Page 23: Chimp child-rearing rates: *Parental Behaviour*, edited by W. Sluckin and M. Herbert (Oxford: Basil Blackwell, 1986).

Page 26: Carnivorous creatures: *Adventures with the Missing Link* by Raymond Dart (Philadelphia: The Institutes Press, 1967). Also see *African Genesis* by Robert Ardrey (New York: Atheneum, 1961).

Page 26: Food gathering and food sharing: *The !Kung of San: Men, Women and Work in a Foraging Society* by Richard B. Lee (Cambridge: Cambridge University Press, 1979); *Stone Age Economics* by Marshall Sahlins (Chicago: Aldine, 1972); "The Food-Sharing Behavior of Protohuman Hominids," Isaac Glynn, *Scientific American*, pp. 90–108 (April 1976).

Page 27: Gorillas can identify more than one hundred types of plant: *The Mountain Gorilla: Ecology and Behavior* by G. B. Schaller (Chicago: University of Chicago Press, 1963).

Page 28: Chimp troops split up to feed: "The Social Ecology of Chimpanzees," Michael Ghiglieri, *Scientific American* 252 (6), pp. 84–93 (1985).

Page 28: Baboon hunt: *The Chimpanzees of Gombe* by Jane Goodall (Cambridge, Mass.: Harvard University Press, 1986); *The Predatory Behavior of Wild Chimpanzees* by G. Teleki (Lewisburg, Pa.: Bucknell University Press, 1973).

Page 29: Hiding food: *The Forest People* by Colin Turnbull (New York: Simon and Schuster, 1961).

Page 30: Breeding patterns: *Primate Societies*, edited by Barbara Smuts (Chicago: Chicago University Press, 1987).

Page 31: Chimps and humans: The question of when man last shared a comon ancestor with the chimps and gorillas has been rather controversial. "Bone gatherers" such as Richard Leakey have argued for a split at least ten million years ago, but recent DNA comparisons seem to give convincing evidence of a much closer relationship between man and the chimps. Try *The Monkey Puzzle*, Gribbin and Cherfas, for a general read; also *The Making of Mankind* by Richard Leakey (New York: Dutton, 1981) and "Evolution at Two Levels in Humans and Chimpanzees," Marie-Claire King and Allan Wilson, *Science* 188, pp. 107–116 (1975).

Page 32: Gene switches: *Ontogeny and Phylogeny* by Stephen Jay Gould (Cambridge, Mass.: Harvard University Press, 1977).

Page 34: Brain size: For the brain sizes of fossil hominids, plus an illustrated guide to the many apes of the Miocene, see *Prehistoric Man* by Vratislav Mazak (London: Hamlyn, 1980). Also "The Casts of Fossil Hominid Brains," Ralph Holloway, in *Human Ancestors: Readings from Scientific American* (San Francisco: W. H. Freeman, 1979).

Page 36: Baby brain growth: "Why Mammals Are Not Bird Brained," Peter Bennett and Paul Harvey, *New Scientist*, pp. 16–17 (April 4, 1985). Also *Ontogeny and Phylogeny*, Gould.

Page 37: Myelinization and brain development: For a general account, see *The Tangled Wing* by Melvin Konner (New York: Holt, Rinehart and Winston, 1982). Also "Plasticity in Brain Development," Chiye Aoki and Philip Siekevitz, *Scientific American*, pp. 34–42 (December 1988).

Page 38: Kitten vision: "Development of the Brain Depends on the Visual Environment," Colin Blakemore and G. F. Cooper, *Nature* 228, pp. 477–478 (1970).

Page 38: Bandaging a child's eyes: An excellent illustrated guide to vision and the structure of the brain is *The Amazing Brain* by Robert Ornstein and Richard F. Thompson (Boston: Houghton Mifflin, 1984).

Page 40: Wild children: in Uganda: "Jungle Boy," article in *The Mail on Sunday*, pp. 32–33 (London, October 11, 1987). Also *Wolf Children and Feral Man* by J.A.L. Singh and R. M. Zingg (New York: Harper, 1942) and *Genie: A Psycholinguistic Study of a Modern-Day "Wild Child"* by S. Curtiss (New York: Academic Press, 1977).

Page 41: Animal tool use: "What's So Special About Using Tools?," Michael Hansell, *New Scientist*, pp. 54–56 (January 8, 1987).

Page 41: Chimp tool use: "Chimpanzees of the Gombe Stream Reserve," Jane Goodall, in *Primate Behavior: Field Studies of Monkeys and Apes*, edited by I. DeVore (New York: Holt, Rinehart and Winston, 1965). Also *The Chimpanzees of Gombe*, Goodall.

Page 42: Potato-washing monkey: "New Acquired Precultural Behaviour of the Natural Troop of Japanese Monkeys on Koshima Inlet," M. Kawai, *Primates* 6, pp. 1–30 (1965).

Page 43: Hominid toolmaking: *Lucy*, Johanson and Edey; *Man The Toolmaker* by K. P. Oakley (Chicago: University of Chicago Press, 1957); "Tools and Human Evolution," Sherwood Washburn, *Scientific American* 203, pp. 3–15 (1960).

Page 46: Big-game hunter: *Early Man* by F. Clark Howell (Chicago: Time-Life, 1973).

Page 47: Rise of *Homo Sapiens: Evolution of the Genus* Homo by William Howells (Reading, Mass.: Addison-Wesley, 1973).

2. The Fisherman's Net

Page 51: Fleeting patterns: For good general introductions on how the brain works: *The Amazing Brain* by Robert Ornstein and Richard F. Thompson (Boston: Houghton Mifflin, 1984); *The Mechanisms of Mind* by Edward de Bono (New York: Simon and Schuster, 1969); *The Fabric of Mind* by Richard Bergland (London: Viking, 1985); *Neuronal Man* by Jean-Pierre Changeux (New York: Pantheon, 1985).

Page 52: Contrast lines: "Mach" lines (named after the discoverer) are discussed in *Visual Perception* by T. N. Cornsweet (New York: Academic Press, 1970). See also *The Handbook of Perception and Human Performance*, Vol. 1, edited by K. Boff, L. Kaufman, and J. P. Thomas (New York: Wiley, 1986).

Page 53: Tricks of vision: For an easy read, see *Odd Perceptions* by Richard Gregory (New York: Methuen, 1986). Also see *Eye, Brain and Vision* by Dave Hubel (New York: W. H. Freeman, 1988) and "Hidden Visual Processes," Jeremy Wolfe, *Scientific American*, pp. 72–85 (February 1983).

Page 54: The pin-sharp fovea: An interesting point about the fovea is that only the primates, among all the mammals, have this supersensitive center to vision. Many reptiles, birds, and fish also have foveas—in fact, birds like the hawk have far better foveas than humans, with cells packed eight times as densely—and moreover, hawks have two foveas—one pointing forward and one down—suggesting they must experience two sharp points of focus at the same time. See *Animal Thought* by Stephen Walker, pp. 269–270 (London: Routledge & Kegan Paul, 1985).

Page 55: The cortex: *The Amazing Brain*, Ornstein and Thompson.

Page 55: Skin sensation mapping: *Biological and Biochemical Bases of Behavior*, edited by H. F. Harlow and C. N. Woolsey, pp. 63–81 (Madison: University of Wisconsin Press, 1965); *Neurobiology* by G. M. Shepherd (Oxford: Oxford University Press, 1983).

Page 56: Visual mapping: *The Amazing Brain*, Ornstein and Thompson. Also "The Discovery of the Visual Cortex," Mitchell Glickstein, *Scientific American*, pp. 84–91 (September 1988), and *Seeing* by J. P. Frisby (New York: Oxford University Press, 1979).

Page 58: Single-cell response to hand and face: "Infero-temporal Cortex and Vision: A Single Unit Analysis," C. Gross, D. Bender, and C. Rocha-Miranda, in *The Neurosciences: Third Study Program*, edited by F. O. Schmidt and F. G. Wordon (Cambridge, Mass.: MIT Press, 1974). Also "The Neural Basis of Stimulus Equivalence Across Retinal Translation," C. Gross and M. Mishkin, in *Lateralization of the Nervous System*, edited by S. Harnad, L. Goldstein, J. Jaynes, and G. Krauthamer (New York: Academic Press, 1977).

It should be noted that some people have talked about these single-cell responses as if the firing of one high-level cell will re-create the whole sensation of a hand or whatever—the so-called grandmother-cell hypothesis. What is argued here is that this grandmother cell is just the tip of the iceberg, the sharp peak of awareness. Its triggering will cause us to experience a sensation only if it sends the right messages fanning back down across the visual areas.

Also, the experiments measuring single-cell responses are controversial. The difficulty of working with something as small as a single brain cell makes it hard to repeat experiments or to be absolutely certain about any particular set of results.

Page 60: Words for capturing the web of nerve activity: Words like *script* and *schemata* come from the cognitive processing branch of psychology and are used more to capture the way networks of ideas seem to hang together than to describe the underlying patterns of cell firing in the gray matter. For want of an established term, I have used *net* to try to tie the mental world of thoughts and sensations to the physical world of fleeting nerve activity.

For more on schemata, see *Remembering* by F. C. Bartlett (New York: Cambridge University Press, 1968). For scripts, see *Scripts, Plans, Goals and Understanding: An Inquiry into Human Knowledge Structures* by R. C. Schank (New York: Halsted Press, 1977).

For more on the general argument that consciousness is the result of webs of nerve activity, see *The Organization of Behavior* by D. O. Hebb (New York: Wiley, 1949).

Page 62: Nets gliding across cortex: Scientists do not yet have sensitive enough instruments to pick out a coordinated wave of cell firing as it spreads across a living brain, but they are making rapid progress from the days when all they could do was tape an electrode to the skull and get a crude message of general brain activity. Recent advances in techniques, such as PET scans (where radioactive sugar is injected into the brain and a computer-driven scanner "films" the uptake of the sugar by hungry brain cells), promise to tell us a lot more about how sensations are mapped out across the cortex.

Page 62: Waxlike rivulets: An excellent description of how the brain surface works, using everyday analogies, is in *The Mechanism of Mind*, de Bono.

Page 63: Nerves sprout new branches: A lot of research has been done recently into rats brought up in "interesting" cages full of toys. The brains of these rats were far richer in connections than those of rats brought up in ordinary laboratory cages. See "Brain Changes in Response to Experience," M. Rosenzweig, E. Bennett, and M. Diamond, *Scientific American* 226, pp. 22–29 (1972).

Page 64: Fleshing out perception with memories: *Cognitive Psychology* by Ulric Neisser (New York: Appleton-Century-Crofts, 1967).

Page 64: Well-trampled visual cortex: The suggestion that the postage stamp of the primary visual cortex carries too heavy a traffic to form any permanent connections is my own speculation. I have found no research that seriously tests the idea. Likewise, there is little hard evidence for the suggestion that all sensation gets funneled down to a few "mother" cells to etch a specific memory in a "memory zone." What evidence there is comes from a number of scattered observations such as the hand-recognition cell (Gross *et al.*). Also, a brain surgeon, Wilder Penfield, has reported on patients who had memories and sensations triggered when he was probing in their brains although Penfield stimulated whole clumps of cells rather than any single cell (*The Mystery of the Mind: A Critical Study of Consciousness and the Human Brain* by W. Penfield (Princeton, N.J.: Princeton University Press, 1975).

On the flip side of the coin, brain damage by blood clots can also be revealing. A recent report (*Nature* 316, p. 388) tells of a stroke victim who lost his ability to name fruit, as if the clot had destroyed the specific clump of cells devoted to the task.

A further point is that the formation of memory traces certainly involves parts of the brain other than the cortex. The hippocampus and other midbrain structures clearly have key roles—if only in channeling and flagging sensations that need to be fixed. For a fuller view of the brain's pathways: *Brain and Mind*, edited by David Oakley (New York: Methuen, 1985). Also "The Anatomy of Memory," Mortimer Mishkin and Tim Appenzeller, *Scientific American*, pp. 62–71 (June 1987).

Page 68: Tack new bits on: For an explanation of how the whole brain was built like a ramshackle house, with new rooms and extensions tacked on to the existing foundations, see *The Amazing Brain*, Ornstein and Thompson, and *Animal Thought*, Walker. Also *Evolution of the Brain and Intelligence* by H. Jerison (New York: Academic Press, 1973).

Page 69: A roomful of experimental gear: "Perceiving a Stable Environment," Hans Wallach, *Scientific American*, pp. 92–98 (May 1985).

Page 70: Special upside-down glasses: The 1920s psychologist G. M. Stratton was the first to try this. See *Eye and Brain* by Richard Gregory (New York: McGraw-Hill, 1966). Also "Experiments with Goggles," Ivo Kohler, *Scientific American* 206 (5), p. 62 (1962), for similar experiments using colored lenses.

Page 74: Slide-recognition experiment: "How We Remember What We See," Ralph Haber, *Scientific American* 222 (5), p. 104 (1970). Also *Recognition and Recall*, edited by J. Brown (London: Wiley, 1976).

Page 76: The "aha" feeling or shock of recognition: Recognition is rarely studied on its own and is usually treated as one of the less exciting aspects of memory. Likewise, little attention appears to have been paid in the literature to the consciously experienced sensation that accompanies recognition. But there is a discussion of "metamemory" in *Memory: A Cognitive Approach* by Gillian Cohen, Michael Eysenck, and Martin LeVoi (Philadelphia: Taylor and Francis, Open University Press, 1986).

For more general accounts of memory: *Your Memory: A User's Guide* by Alan Baddeley (New York: Macmillan, 1982); *Memory in the Real World* by Gillian Cohen (Hillsdale, N.J.: Lawrence Erlbaum Associates, 1989); and *Memory Observed: Remembering in Natural Contexts*, edited by Ulric Neisser (San Francisco: W. H. Freeman, 1982).

Page 77: Déjà vu: This is my own explanation, based on what we know about the recognition feeling and metamemory. For other theories, see "Temporal Perception, Aphasia and Déjà Vu," R. Efron, *Brain* 86, p. 403 (1963).

Page 79: Memory traces left behind by firing nets: The location of memory traces has been one of the most widely speculated upon points in memory research. Suggestions have ranged from organic holograms (*Brain and Mind*, ed. Oakley, pp. 59–98) to the encoding of memories on proteins stored inside neurons (based on experiments such as "Memory in Mammals," D. J. Albert, *Neuropsychologia* 4, pp. 79–92, 1966). But recent work on the sea slug, *Aplysia* (referred to in detail in Chapter 3), points to memory as being the result of the forging of new dendrite connections and sensitivity changes at nerve-junction membranes.

3. Rousing Memories

Page 86: Tablecloth of human cortex: *Animal Thought* by Stephen Walker (New York: Methuen, 1985).

Page 86: Tests of animal memory: "Cognitive Mapping in Chimpanzees," E. W. Menzel, in Cognitive Processes in Animal Behavior, edited by S. H. Hulse, H. Fowler and W. K. Honig (Hillsdale, N.J.: Lawrence Erlbaum Associates, 1978); Vertebrate Memory, J. S. Beritoff (New York: Plenum Press, 1971); "Memory in Food Hoarding Birds," Sara Shettleworth, Scientific American 248 (3), pp. 86–94 (1983). Also, a general treatment in Animal Thought, Walker, pp. 287–338.

Page 87: Hardware of human brain is not fundamentally different from that of other creatures: See Animal Thought, Walker, pp. 142–193.

Page 88: Language as the dividing line: The observation that language is the key ability separating man from the animal seems quite straightforward, although over the centuries philosophers like Descartes have tended to look for spiritual or metaphysical differences between animals and humans. Hume, in the 1700s, was about the first to argue clearly the importance of language but his arguments were overshadowed by his more conservative colleagues for over a hundred years until Schopenhauer picked up the thread again in the late 1800s (The World as Will and Idea by A. Schopenhauer).

In this century it has often been remarked by psychologists that speech is the dividing line (see Sociobiology by E. O. Wilson, p. 556, Cambridge, Mass.: Harvard University Press, 1975), but this does not appear to have led to a systematic analysis of exactly how language makes man's mind different. One subject of much experimentation, however, is whether chimpanzees can be taught to speak and so develop humanlike minds. Researchers have realized that chimps lack the necessary vocal tract and brain centers ever to manage the rapid-fire speaking of humans, but they have got somewhere teaching them to use symbols to communicate. There has been no evidence that this has led to a symbol-driven thought mimicking what goes on in the human mind—and this would not be expected; as this book argues, humans needed a rapid inner voice, hundreds of thousands of years of cultural evolution, and an intense social upbringing to create an inner driver for the mind. To see how far chimps could progress, try Nim by H. S. Terrace (New York: Knopf, 1979) and Animal Thought, Walker, pp. 352–381.

Page 90: Remembering lists and drinks orders: "How Big Is a Chunk," H. A. Simon, Science 183, pp. 482–488 (1974); "Remembering Drinks Orders: The Memory Skills of Cocktail Waitresses," Henry Bennett, Human Learning 2, pp. 157–169 (1983).

Page 91: Brain damage and lack of long-term memory: A moving documentary (Prisoner of Consciousness, Equinox, Channel Four, London, 1987) was made about such a victim of blood clot. See also The Clinical Management of Memory Problems by Barbara Wilson (Rockville, Md.: Aspen Publications, 1984).

Page 94: Eyes caught by sudden movements: There is good evidence that a special bulge of neurons in the brain acts as an alerting center for our sense of vision. This midbrain center, like the cerebellum, is intelligent but does not give rise to consciously experienced sensations, which leads to an uncanny phenomenon known as blind sight. When stroke victims whose visual cortex had been destroyed and who thus believed they were completely blind were asked to point to a lighted bulb on a grid of lights that the researchers could turn on at random, they protested that they could see nothing yet they still pointed straight at the glowing bulb. Their still intact midbrain did the seeing—even if such sight was unconscious and able to spot only sudden movements or changes in illumination. The ability was first observed in chimps that had had their cortexes removed. See "Vision in a Monkey Without Striate Cortex: A Case Study," N. Humphrey, *Perception* 3 (3), pp. 241–255 (1974).

The midbrain center responsible for this blind sight happens to be the primary visual area for reptiles and birds, whose midbrains have become enlarged to form what is known as the optic tectum. At some time in their evolution from the reptiles, the mammals transferred the job of mapping vision to the cerebrum, which at that stage was only a small bulge of brain for handling the olfactory senses of taste and smell (*Animal Thought*, Walker).

Page 94: Noises start out on the fringe of awareness and then thrust themselves into the center of attention: One dramatic example is the cocktail-party phenomenon, when we can be talking to someone and suddenly, against the background buzz of noise, latch on to a person mentioning our name or saying something outrageous. We must have been monitoring the hubbub at some level of intelligence in order to pick up so quickly on something interesting. The likely explanation for this ability is that all the conversations stayed on the fringes of awareness until one phrase struck a chord of recognition and the "aha" response jerked our attention toward that distant conversation. The same sort of thing happens when we are driving and relying on our intelligent fringes to cope with the routine and call on our full attention only in an alarming or difficult situation.

For experiments on the cocktail-party phenomenon, see "Attention in Dichotic Listening: Affective Cues and the Influence of Instructions," N. Moray, *Quarterly Journal of Experimental Psychology* 11, pp. 56–60 (1959).

Page 95: Associative memory forearms us with knowledge: Many studies have been done to show how much faster and more easily we can recall something (usually a word or a picture in psychological tests) if a closely related sound or idea has recently been awakened. See "Associative Encoding and Retrieval: Weak and Strong Cues," D. M. Thomson and E. Tulving, *Journal of Experimental Psychology* 86, pp. 255–262 (1970), and "Structure of Memory Traces," E. Tulving and M. I. Watkins, *Psychological Review* 82, pp. 261–275 (1975).

For the importance of being forearmed, see *Behavioural Ecology: An Evolutionary Approach*, edited by J. R. Krebs and N. B. Davies (Sunderland, Mass.: Sinauer Associates, 1978).

Page 96: Shereshevskii's memory: *The Mind of a Mnemonist* by A. R. Luria (New York: Basic Books, 1968).

Page 98: Photographic memory, Nigerians and Western children: "Eidetic Images Among the Ibo," L. W. Doob, *Ethnology* 3, pp. 357–363 (1964); "Eidetic Imagery: Frequency," R. H. and R. B. Haber, *Perceptual and Motor Skills* 19, pp. 131–138 (1964).

Page 100: Learning in *Aplysia*, the sea slug: For a general introduction, *Explorers of the Black Box: The Search for the Cellular Basis of Memory* by Susan Allport (New York: W. W. Norton, 1987).

Page 103: This simple rewiring is the basis of all learning and memory in higher animals: The fixing of long-term memories is undoubtedly a more complex process than there has been room to outline in this book. Special parts of the brain, like the hippocampus, appear to play an important role in what is stored and how it is retrieved. Neurologists have still to uncover much of the story, but the simple principle of rewiring appears to be at the heart of the process.

Page 104: Chemical changes at the synapse: *Synapses, Circuits and the Beginnings of Memory* by Gary Lynch (Cambridge, Mass.: MIT Press, 1986).

Page 105: Unique images and blurred nets: The existence of snapshotlike memory is often remarked upon but the blurred nature of most memories, and the way they need to be re-created, are not. See, for example, *Autobiographical Memory*, edited by David Rabin (New York: Cambridge University Press, 1987). Also *Your Memory: A User's Guide* by Allan Baddeley (New York: Macmillan, 1982). Or, for a criticism of so-called flashbulb memories, "Vivid Memories," D. C. Rabin and M. Kozin, *Cognition* 16, pp. 81–85 (1984).

The mechanism proposed in this book for both types of memory net is mostly my own speculation.

Page 106: The importance of a general world picture: Recent research into artificial intelligence in computers has brought this into focus. See *Machines Who Think* by Pamela McCorduck (San Francisco: W. H. Freeman, 1979) and *Issues in Cognitive Modeling*, edited by A. M. Aitkenhead and J. M. Slack (Hillsdale, N.J.: Lawrence Erlbaum Associates, 1985). Also "Intuitive Physics," Michael McCloskey, *Scientific American* 248 (4), pp. 114–122 (1983).

Page 108: Recall strategies to jog our memories: This inner dialogue to dredge up memories is not much studied, largely because it would require an unfashionable amount of introspection by the experimental subject. But see "Transformations of Memory in Everyday Use," M. Linton, in *Memory Observed*, edited by Ulric Neisser (San Francisco: W. H. Freeman, 1982), and "Directed Search Through Autobiographical Memory," W. B. Whitten and J. M. Leonard, *Memory and Cognition* 9, pp. 566–579 (1981).

Page 1100: Words are only noises: *Issues in Cognitive Modeling*, eds. Aitkenhead and Slack, pp. 159–206, and "Slips of the Tongue," Michael Motley, *Scientific American* 253 (3), pp. 114–119 (1985).

Page 110: Naming something makes it stand out: The eighteenth-century philosophers Kant and Hegel were the first to make something of the effect words have on our perception of the world, a vein of thought that is still enthusiastically pursued today by philosophical schools such as structuralism and semiotics. Some psychologists have also picked up on it, the classic text being *Language, Thought and Reality* by B. L. Whorf (Cambridge, Mass.: Technology Press, 1956). The tendency has often been to go too far and argue that our words effectively create the world we perceive. Some sobering research, such as a test of New Guinea tribesmen who have no words for colors yet can still tell two shades of red apart as readily as any Westerner, shows the danger of mistaking words to be more than handy labels for the perceptions themselves (see "On the Internal Structure of Perceptual and Semantic Categories," E. Rosch, in *Cognitive Development and the Acquisition of Language,* edited by T. E. Moore, New York: Academic Press, 1973).

Page 114: Top-level cells sparking to life to re-create images: This is largely my own speculation about what happens, but support can be drawn from studies that show that electrical stimulation of certain parts of the brain can prompt detailed images and sounds in the minds of brain-surgery patients. See *The Mystery of the Mind: A Critical Study of Consciousness and the Human Brain* by W. Penfield (Princeton, N.J.: Princeton University Press, 1975); "Mental Imagery and the Visual System," Ronald Finke, *Scientific American* 254 (3), pp. 76–83 (1987); "Concerning Imagery," D. O. Hebb, *Psychological Review* 75, pp. 466–477 (1968).

Page 118: Eyewitness evidence at trials: *Eyewitness Testimony* by E. F. Loftus (Cambridge, Mass.: Harvard University Press, 1979).

Page 119: Factual memory as another form of re-created memory: Factual (or semantic) memory is usually treated as if it were recalled directly and verbally, without any visual or re-created component (see, for example, *Your Memory: A User's Guide,* Baddeley). However, a little unfashionable introspection quickly shows that strategies for recalling semantic memories are much like those for personal (or episodic) memories.

Page 121: Self-consciousness is clearly based on memory: This does not appear to be a widely understood or much appreciated distinction, although it has often been remarked upon in passing. Cognitive scientists, with their "re-representation of representation systems" (i.e., memories of being aware), seem to come the closest to incorporating this key point in their theories. See, for example, *Brain and Mind,* edited by David Oakley, pp. 217–251 (New York: Methuen, 1985).

Page 122: The gap between doing and reviewing the doing in hitting a tennis ball: The tennis instruction book, *The Inner Game of Tennis* by Timothy Gallway (New York: Bantam, 1979) captures this well.

4. Thinking Aloud

Page 126: Chains of ideas: This view of how nets interact is modeled on the way nerve cells behave on the retina. Also, for a more metaphorical explanation of thinking, see *The Mechanism of Mind* by Edward de Bono (New York: Simon and Schuster, 1970).

Page 128: Cortex complexity and inner representations: *Animal Thought* by Stephen Walker (New York: Methuen, 1985).

Page 129: Dolphin mimics: "Imitative Behaviour in Indian Ocean Bottlenose Dolphins in Captivity," C. K. Tayler and G. S. Saayman, *Behviour* 44, pp. 286–298 (1973).

Page 130: Chimps reaching for a banana: An experiment repeated with many variations but based on classic work by W. Kohler in *The Mentality of Apes* (London: Kegan Paul, Trench and Trubner, 1925).

Page 130: Children reaching with clamps and sticks: *Play: Its Role in Development and Evolution*, Chapter 24, edited by J. Bruner, A. Jolly, and K. Sylva (New York: Basic Books, 1976).

Page 131: A dog cannot be afraid its master will beat it tomorrow: *Philosophical Investigations*, para. 650, Ludwig Wittgenstein (Oxford: Basil Blackwell, 1976).

Page 133: Babies have to learn that things still exist when hidden from sight: *Developing Thinking*, edited by S. Meadows (New York: Methuen, 1983).

Page 134: Truk Islander navigation: "Culture and Logical Progress," T. Gladwin, in *Explorations in Cultural Anthropology*, edited by W. H. Goodenough (New York: McGraw-Hill, 1964).

Page 137: Puzzle of treating stomach cancer: The original problem comes from "Analogic Problem Solving," M. L. Gick and K. J. Holyoak, *Cognitive Psychology* 12, pp. 306–356 (1980). The interpretation in terms of nets is my own speculation.

Page 139: We do not understand things unless they are grounded in everyday experience: For a general text, see *Metaphors We Live By* by George Lakoff and Mark Johnson (Chicago: University of Chicago Press, 1980).

Page 140: Dominant metaphors of cultures: A readable treatment, concentrating on the latest, the computer, is in *Turing's Man* by J. David Bolter (Chapel Hill, N.C.: North Carolina University Press, 1984).

Page 140: Freud's hydraulic view of human nature: Freud—along with Jung, Adler, and Erikson—developed what have come to be known as psychodynamic or depth theories of human personality. The hydraulic metaphor behind these theories is clear from the way Freud talked about repression and the unconscious when he first outlined his ideas in *The Interpretation of Dreams*.

5. From Baby Talk to Strong Language

Page 143: Rise of *Homo sapiens:* A crop of modern skulls dating from this time has been discovered in Europe and given the collective name of Cro-Magnon from the French cave where the key find was made. This has led to a depiction of *Homo sapiens* as an intelligent white European who quickly replaced the brutish Neanderthals, but older and more widespread finds have recently made it seem more likely that *Homo sapiens* first arose in North Africa or the Middle East, spreading into the snowy wastes of Europe—the natural habitat of sturdier Neanderthals—only after being established for some tens of thousands of years. See *Evolution of the Genus Homo* by William Howells (Reading, Mass.: Addison-Wesley, 1973) and *Paleoanthropology* by G. E. Kennedy (New York: McGraw-Hill, 1980).

Page 143: Early theories on evolution of language: A good general text on language is *The Story of Speech and Language* by C. L. Barber (New York: Apollo, 1972). Also *Language Origins: A Bibliography* by G. Hewes (Atlantic Highlands, N.J.: Humanities, 1975).

Page 144: Vervet monkey calls: "How to Listen to the Animals," Robin Dunbar, *New Scientist,* pp. 36–39 (June 13, 1985).

Page 145: Range of chimp and human vocalizations: For chimps, see *On the Origins of Language* by P. Lieberman (New York: Macmillan, 1975). The variety of human vocalization can be seen in the many different syllabic sounds used in human languages. English, for example, uses forty-five separate speech sounds (while Hawaiian manages on about thirteen). See *The Story of Language,* Barber, pp. 2–15.

Page 146: Angry hooting to put off bold young male: Described in the excellent book on chimp behavior at Arnhem Zoo, *Chimpanzee Politics* by Frans de Waal (London: Jonathan Cape, 1982).

Page 147: Social animals need to be able to model mentally the social world of their group: "Cognitive Mapping in Chimpanzees," E. W. Menzel, in *Cognitive Processes in Animal Behavior,* edited by S. H. Hulse, H. Fowler, and W. K. Honig (Hillsdale, N.J.: Lawrence Erlbaum Associates, 1978). Also "Nature's Psychologists," N. K. Humphrey, in *Consciousness and the Physical World,* edited by B. D. Josephson and V. S. Ramachandran (New York: Pergamon Press, 1980).

Page 152: Chimp murders at Gombe Stream Reserve: *The Chimpanzees of Gombe* by Jane Goodall (Cambridge, Mass.: Harvard University Press, 1986); "Life and Death at Gombe," Jane Goodall, *National Geographic,* pp. 592–621 (May 1979).

Page 153: Attention-directing games between mother and child: "The Development of Naming," John McShane, *Linguistics* 17, pp. 879–905 (1979).

Page 156: The first words arose accidentally: This account of the evolution of language is largely my own speculation. The gestures used by chimps come from *Chimpanzee Politics,* de Waal. Other influential works: *Biolog-*

ical Foundations of Language by Eric Lenneberg (New York: Wiley, 1967); "The Origin of Speech," Charles Hockett, *Scientific American* 203, pp. 88–111 (1960).

Page 159: Evolution of the human vocal tract: *The Biology and Evolution of Language* by Philip Lieberman (Cambridge, Mass.: Harvard University Press, 1984); "The Anatomy of Human Speech," Jeffrey T. Laitman, *Natural History* (August 1984); *Glossogenetics: The Origin and Evolution of Language,* edited by Eric de Groller (New York: Harwood Academic Publishers, 1981).

Page 162: Is speech hard-wired or learned?: The consensus is that it is a bit of both, but under the powerful influence of Noam Chomsky, the weight of scientific opinion leans more toward a hard-wired or genetically programmed speech acquisition mechanism than I do in this book—especially as far as grammar is concerned. For more on this controversy, try *Language Development,* edited by Andrew Lock and Eunice Fisher (London: Croom Helm, 1984); *Language and Mind* by Noam Chomsky (New York: Harcourt Brace Jovanovich, 1968); "Creole Languages," Derek Bickerton, *Scientific American* 249 (1), pp. 108–115 (1983).

Page 163: Visible swellings of the speech centers: "Right-Left Asymmetries in the Brain," A. Galaburda et al., *Science* 199, pp. 853–856 (1978).

Page 163: Stroke damage and reading problems: *The Tangled Wing* by Melvin Konner (New York: Holt, Rinehart and Winston, 1982), p. 159.

Page 167: Evolution wired up *Homo erectus* to make gestures: The need for a lot of explanatory gestures to aid comprehension of speech in its early days may be the answer to a puzzling phenomenon: We still use a lot of "body language" when we talk today, although most humans do not seem to pay much attention to these extra clues. See *Manwatching* by Desmond Morris (New York: Abrams, 1977) and *Body Language* by J. Fast (New York: Evans, 1970).

Page 169: Hopi and Navaho Indian grammar: "A Systematization of the Whorfian Hypothesis," J. A. Fishman, *Behavioral Science* 5 (4), pp. 323–339 (1960).

Page 171: Neanderthal culture: *The Old Stone Age* by F. Bordes (New York: McGraw-Hill, 1968); *Shanidar: The First Flower People* by Ralph Solecki (New York: Knopf, 1971).

Page 172: Explosion of *Homo sapiens* culture: "Visual Thinking in the Ice Age," Randall White, *Scientific American,* pp. 74–81 (July 1989).

Page 172: Only about ten million *Homo sapiens: Atlas of World Population History* by C. McEvedy and R. Jones (New York: Penguin, 1978).

Page 173: Domestication and cultivation: *The Neolithic Revolution* by S. Cole (London: British Museum [National History], 1970).

6. Strange Voices in the Head

Page 176: The inner voice: This is a little studied part of the mind, presumably because it would take an unfashionable amount of introspection to document. An otherwise comprehensive book like *Language Development*, eds. Andrew Lock and Eunice Fisher (London: Croom Helm, 1984) makes no mention of it. This is typical.

One of the few researchers to give the inner voice any attention has been Lev Vygotsky in *Thought and Language* (Cambridge, Mass.: MIT Press, 1962). For a review of Vygotsky's work on the inner voice and also the social shaping of self-awareness, see *Vygotsky and the Social Formation of the Mind* by James Wertsch (Cambridge, Mass.: Harvard University Press, 1985).

Page 177: Foreground and background of awareness: Again, very little work has been done to portray accurately the feeling of being aware, even though the essence of a fringe and sharp focus was noted as long ago as 1890 (W. James, reprinted in *The Principles of Psychology*, London: Constable, 1950).

Page 178: Repeating a monotonous phrase to blot out the inner voice: The mind-numbing chants and mantras of many religions are clearly trying to achieve the same effect of quelling the inner voice and "stilling" the mind. Perhaps with practice it is possible to do so. See *Altered States of Consciousness: A Book of Readings*, edited by C. T. Tart (New York: Wiley, 1969). Also *Essentials of Zen Buddhism* by D. T. Suzuki (New York: Dutton, 1961).

Page 181: Adults as language teachers: Much has recently been written about how adults cocoon babies in language lessons. See "Conversations with Children," Catherine Snow, in *Language Acquisition*, edited by R. Fletcher and M. Garman (New York: Cambridge University Press, 1979). Also "How Children Learn Words," G. Millar and P. Gildea, *Scientific American* 257 (3), pp. 86–91 (1987).

Page 182: The stages in learning speech: *Language Development*, Lock and Fisher. Another good general guide is *Baby Language* by Maire Messenger Davies, Eva Lloyd, and Andreas Scheffler (London: Unwin Paperbacks, 1987).

Page 185: Thinking aloud in children: See *Thought and Language*, Vygotsky. Also *How Children Think and Learn* by David Wood (Oxford: Basil Blackwell, 1988); "Development of Private Speech Among Low Income Appalachian Children," Laura Berk and Ruth Garvin, *Developmental Psychology* 20 (2), pp. 271–286.

Page 187: Deaf people and speech through gestures: Given the right conditions, gesture language can flourish and even evolve. See *Everyone Here Spoke Sign Language: Hereditary Deafness on Martha's Vineyard* by Nora Ellen Groce (Cambridge, Mass.: Harvard University Press, 1987). For a discussion of the effects of not having speech, see *Thinking Without Language* by Hans Furth (New York: Free Press, 1966).

Page 193: Feats of memories in earlier times: *The Art of Memory* by Frances Yates (Chicago: University of Chicago Press, 1966).

Page 196: Conscience: This interpretation of conscience as depending on the imagined condemnation and praise of our peers is largely my own. But see *Moral Development and Behavior: Theory, Research and Social Issues*, edited by T. Lickona (New York: Holt, Rinehart & Winston, 1976). Also *Psychology: The Science of Mind and Behaviour* by Richard Gross, pp. 524–554 (London: Edward Arnold, 1987) and *Morality in the Making* by Helen Weinreich-Haste and Don Locke (New York: Wiley, 1973).

Page 199: We feel the weight of society's expectations only if we swim against the tide: There have been many dramatic experiments showing how bystanders in public places ignore people collapsing with heart attacks or being mugged. Everyone is frozen because no one can bring himself to make the first step. See *The Unresponsive Bystander: Why Doesn't He Help?*, by B. Lantane and J. M. Darley (New York: Appleton-Century-Crofts, 1970), where in one of their experiments it was shown that the herd instinct rules even with a Frisbee game played in the waiting room of a station. Two girls threw a Frisbee back and forth for a while, then "accidentally" threw it in the direction of a confederate. If the confederate happily slung it back, most of the rest of the people in the room would join in the game, but if the confederate kicked it back and scolded the girls, no one else would join in either.

For a more general discussion of the weight of society's expectations, see *The Approval Motive* by Douglas Crowne and David Marlowe (New York: Wiley, 1964). For more on individuality and free will, see *The Determinants of Free Will* by James Easterbrook (New York: Academic Press, 1978).

Page 199: Simple primitive societies with no conception of other life-styles: A well-written illustration is to be found in the difficulty that the Congo pygmies in the *Forest People* (Colin Turnbull, New York: Simon and Schuster, 1961) had understanding the customs of the local village Africans, let alone white Westerners.

Page 201: Social roles and a bagful of masks: *Social Interaction and Its Management* by Judy Gahagan (New York: Methuen, 1984).

Page 204: Variety of modern languages: *The Story of Language* by C. L. Barber (New York: Apollo, 1972).

Page 204: Early writing and math: "Numbers and Measures in the Earliest Written Record," Joran Friberg, *Scientific American* 250 (2), pp. 78–85 (1984). Also *The Story of Language*, Barber.

Page 205: Learning to read silently: *The Day the Universe Changed* by James Burke, pp. 101–102 (London: BBC Books, 1985).

Page 208: Cheaper and easier to rediscover than to find academic texts: This is of course the one quote where I did not note the source—and now cannot track it down!

7. Pure Emotions and Romantic Notions

Page 211: The metabolic advances of mammals and the need to control emotions: *Mesozoic Mammals: The First Two-Thirds of Mammalian History*, edited by J. A. Lillegraven, Z. Keilan-Jaworowska, and W. A. Clemens (Berkeley, Calif.: University of California Press, 1979).

Page 212: Lost under a welter of labels: *Explorations in Personality* by H. A. Murray (New York: Oxford University Press, 1938). Also *The Language of Emotion* by Joel Davitz (New York: Academic Press, 1969).

Page 213: Higher emotions as human inventions: The argument that what we call higher emotions are really cocktails of socially valuable ideas, laced with a "raw" emotion for extra punch, is mostly my own speculation. I have not seen it well articulated elsewhere, although there are plenty of hints and snippets to be found. For a review of theories of emotion, see *The Psychology of Emotion* by K. Strongman (New York: Wiley, 1973).

Page 214: The brain stem and arousal: First noted by G. Moruzzi and H. Magoun, "Brain Stem Reticular Formation and Activation of the Electroencephalogram," *Electroencephalography and Clinical Neurophysiology* 1, pp. 455–473 (1949).

Page 215: Arousal levels felt consciously through many small telltale signs: Everyday language is full of phrases that give evidence of how we are aware of the many metabolic changes that take place as our arousal levels change. For example: "My heart leaped into my mouth," "He was white with shock," "Her mouth went dry." There is also a lot written about arousal but little that spells out the way our conscious feelings of excitement or relaxation are really built up from observations of these body changes.

Page 216: Fear and rage as two sides of the same coin: The way that both fear and rage are based on the same metabolic "panic" systems has been well studied. See *The Psychology of Behavior* by Neil Carlson (Boston: Ally & Bacon, 1979) or *Motivation and Emotion* by Phil Evans (London: Routledge, 1989).

Page 217: Cool-headed aggression of hunting cats versus wild rage of apes: *The Tangled Wing* by Melvin Konner, pp. 183–207 (New York: Holt, Rinehart and Winston, 1982).

Page 218: Classic experiment with adrenaline: "Cognitive, Social and Physiological Determinants of Emotional State," S. Schachter and J. E. Singer, *Psychological Review* 69, pp. 379–399 (1962).

Page 219: Pleasure and pain centers in the brain stem: "Pleasure Centers in the Brain," J. Olds, *Scientific American* 195 (4) (1956); *The Puzzle of Pain* by R. Melzack (New York: Basic, 1973).

Page 221: The chemistry behind moods of happiness and depression: *Motivation and Emotion* by Phil Evans.

Page 222: Pain blockers and runner's high: "Identification of Two Related Pentapeptides from the Brain with Potent Opiate Antagonist Activity," J. Hughes, *Nature* 258, pp. 577–579 (1975).

Page 222: Feelings of thirst promoted by chemical message: "The Way We Act," Yvonne Baskin, *Science* 85, pp. 94–100 (November 1985).

Page 228: The basis of love and romance: For a general look at current approaches to analyzing love, see *The Tangled Wing*, Konner, pp. 291–324. Also "The Nature of Love," H. E. Harlow, *American Psychologist* 13, pp. 673–685 (1958).

8. Watching the Watcher

Page 235: The inner watcher inside our heads fallacy: "The Homunculus Fallacy," A. Kenny, in *Interpretations of Life and Mind*, edited by M. Grene (Atlantic Highlands, N.J.: Humanities, 1971). Also *The Modularity of Mind*, J. Fodor (Cambridge, Mass.: MIT Press, 1982).

Page 236: Self-consciousness as simply an egocentric net of ideas and habits of thought: Introspection has been so out of favor among psychologists that there is only a thin thread of research into the way self-consciousness is the turning of consciousness around to observe itself. The issue was touched upon by F. Bartlett in *Remembering* (Cambridge: Cambridge University Press, 1932) and put more forcefully by G. H. Mead in *Mind, Self and Society* (Chicago: University of Chicago Press, 1934). More recently, with the rise of computer-based models of psychology, it has become respectable at least to speculate about representations and re-representations (see *Brain and Mind*, edited by David Oakley, New York: Methuen, 1985).

It is also worth looking at studies of multiple personalities to see how it seems possible that we can form more than just one knot of self-awareness ideas. See *The Final Face of Eve* by E. Lancaster and J. Poling (New York: McGraw Hill, 1958). And for a look at the sources of information about the self, see *Self-Perception* by Chris Kleinke (San Francisco: W. H. Freeman, 1978).

Page 238: We see things as wholes, not just as the sum of their parts: *Perceptual Organization*, edited by M. Kubovy and J. Pomerantz (Hillsdale, N.J.: Lawrence Erlbaum Associates, 1981). Also "Features and Objects in Visual Processing," Anne Treisman, *Scientific American* 225 (5), pp. 106–115 (1986).

Page 243: The active process we call attending is in fact the final result of the natural self-organization that goes on in the brain: The mechanisms behind attention are poorly understood and much of what I say is speculation, based on general ideas about net behavior. It is also likely to be a considerable oversimplification since so many different parts of the brain have been found to be involved in the attention process.

For more on attention, see *Preconscious Processing* by N. E. Dixon (New York: Wiley, 1981); "A Feature-Integration Theory of Attention," A. M. Treisman and G. Gelade, *Cognitive Psychology* 12, pp. 97–136 (1980); and "Brain Mechanisms of Visual Attention," R. Wurtz, M. Goldberg, D. Lee

Robinson, *Scientific American* 246 (6), pp. 100–107 (1982). For a discussion of attention theories, *Brain and Mind*, Oakley, pp. 71–76.

For a look at the dangers of doing well-learned tasks on autopilot, without careful attention, see *Aspects of Consciousness, Vol. I: Psychological Issues*, pp. 67–89, edited by G. Underwood and R. Stevens (New York: Academic Press, 1979).

Page 245: Cerebellum: The cerebellum may lack consciousness—or rather the ability to map sensations—but it is far from unintelligent. At the same time as the cortex of *Homo sapiens* was ballooning, so was the cerebellum—presumably to help tool-using man in the many complex muscular movements he makes with his hands. For a description of the cerebellum's working: "The Cerebellum as a Computer: Patterns in Space and Time," J. C. Eccles, *Journal of Physiology* 229, pp. 1–32 (1973).

Page 249: The unconscious of Freud: The problems with Freud's work have been well documented, but Freud did a lot of well-respected work on the neurology of the brain before going on to expound his famous psychoanalytical theories. These became part of a world of therapy and have often been treated as unquestionable dogma rather than scientific theories to be challenged, tested, and modified, which has led to Freud's being remembered and damned mostly for the excesses of his work.

For a primer on Freud's ideas, see *Psychology and Freudian Theory* by Paul Kline (New York: Methuen, 1984). For criticism of Freud, see "Methodological Issues in Psychoanalytical Theory," E. Nagel, in *Psychoanalysis, Scientific Method and Philosophy*, edited by S. Hook (New York: New York University Press, 1959); also *The Scientific Credibility of Freud's Theories and Therapy* by S. Fisher and R. P. Greenberg (New York: Basic, 1977).

9. Truly Self-conscious Man

Page 259: This is not to say that we will move on to some blissful higher plane: I have not talked in this book about ESP or any other such "paranormal" powers of the mind because I have seen no convincing evidence for such powers. Furthermore, common sense gives more than enough evidence against their existence.

The best argument against the paranormal is to ask what would happen if such powers really existed: If telepathy (thought reading), telekinesis (object shifting), or premonition actually existed, then, for a start, no one could make any money from running gambling games. Slot machines are programmed to give their owners a profit margin calculated to a split percentage and even the slightest distortion of these predetermined pay-out odds by gamblers armed with psychic powers would soon show up in a casino running thousands of one-armed bandits. Likewise, the spin of a roulette wheel, the outcome of horse races, the picks for the football pools—

none of these forms of gambling could survive if even a tiny proportion of the population had supernatural powers. And one can hardly argue that gamblers do not put all their mental energies into willing their number to come up!

While I believe that we can learn to be more understanding of ourselves, more aware of society's influences, and more in control of our own mental development, I do not see any hidden powers waiting to be uncovered.

Index

adrenaline, 217, 218, 227
Africa, 24
    apes of, 16, 17–18
    hominids of, 20, 46, 47
    modern hunter-gatherers of, 26–29
    predators of, 18, 19
aggression, 216, 231
    cold-blooded, 217–218
"aha" feeling, 75–77, 78, 107, 130, 177
alarm calls, 144, 145
alphabet, 205
Amazon jungle Indians, 197
Andaman Islanders, 45
animals, 85–89, 142, 210–211, 219–220, 242
    communication of, 144–146, 186
    consciousness of, 86, 87–89, 121, 188
    domestication of, 26, 173
    emotional cries of, 146, 186
    emotion in, 131–132
    learning by, 41–42, 100–103, 132–133, 187, 189
    memory in, 86–87, 95–96, 100–103, 105–106, 110, 111, 116–117, 122
    perception in, 85, 106, 111
    thought in, 128–130, 131–133, 138
    uncluttered minds of, 88
    see also specific animals
antisocial behavior, 252, 253
    group, 150–151, 201–202
apes, 13, 89
    aggression of, 217
    Asian, 16, 20
    brain size of, 17, 33–34, 37
    evolution of, 15–18, 151
    extinction of, 17
    human genetic closeness to, 31–32, 33, 153, 262
    intelligence of, 17, 24, 25, 26, 145, 146–147, 149, 258
    knuckle-running of, 16, 18
    modern populations of, 17–18
    reproductive strategy of, 17, 22, 23–25, 36, 37, 151
    sexual signals of, 30
    social organization of, 18, 24, 29–30
    species of, 16
    vocal tract of, 159–160
Aplysia (sea slugs), 100–103, 210
arithmetic, 136, 205, 261
Arnhem Zoo, 147–149, 151
arousal, 214–218, 219, 223, 226–227
    biorhythms in, 215
    brain stem in, 214–215, 217, 220
    passions in, 216–218
art, 213, 228
    prehistoric, 172
Asia:
    apes of, 16, 20
    hominids of, 46

association of ideas, 81–82, 127
    in recall, 108, 109–110
    in working memory, 95–96
astronauts, 248
attention, 61, 111, 127, 177–178, 181, 188, 216, 241–249, 256
    concentration of, 243, 246, 260
    distractions to, 241, 246, 247, 260
    divided, 244–245
    focus of, 93–95, 241–242, 244–246
    gaze-following reflex of, 153–154, 162
    nets in, 241, 242–244, 245–247
    shifts of, 246–247
    working memory and, 93–95, 244
Australian aborigines, 26, 43, 171
Australopithecus, 20–21, 22, 32, 46
    brain size of, 34
    extinction of, 34, 44
axes, 42–44

baboons, 19, 27, 30, 41, 87, 189
baby talk, 181
banishment, 197
baskets, 43
biorhythms, 215
bipedalism, 18, 19, 22, 24, 25, 33
birds, 28, 147
    communication of, 145–146
    food memory of, 86, 87
    reproductive strategies of, 34–36
    tool use by, 40–41
Bowwow theory, 143
brain, 85–86, 87–88, 131, 188, 211, 259
    cerebellum of, 245
    cerebral hemispheres of, 54, 55
    cerebrum of, 55
    chemical changes in, 62–63
    cortex of, 55–60, 79, 85, 163–164, 215, 219
    damage to, 40, 90–91, 163–164
    emotions and, 214–216, 217, 219–220, 221–222, 250
    energy used by, 17, 47
    evolution of, see evolution
    limbic system of, 250
    mammalian, 13–14
    of newborn infants, 37, 38–39, 162
    nonconscious parts of, 250
    occipital lobe of, 56
    postnatal development of, 37, 38–40
    prenatal development of, 36–38
    processing zones of, 55–58, 64–65, 69, 75
    relearning by, 40, 164
    self-awareness of, 121–122
    skin sensations and, 55–56
    speech centers of, 39–40, 162–165, 178–184
    thalamus of, 54–55

brain (continued)
  visual cortex of, 56–60, 61, 65, 69, 73, 75, 82, 112
  see also mind, nets, neurons
brain size, 86, 87, 187
  of apes, 17, 33–34, 37
  of hominids, 18, 19, 20, 34, 43, 44, 47, 163, 171
  increase of, in evolution, 13, 19, 25, 31, 33–40, 103
brain stem, 214–215, 217, 220

calpain, 104
campsites, 25, 29, 45, 170, 190
cartoons, 112
cats, 36, 41, 131–132
  cold-blooded aggression of, 217
  consciousness of, 88–89
cave bears, 21, 171
children, 150, 157–158, 181–185, 249, 261
  cultural evolution and, 191–195
  feral, 40
  grammar acquired by, 183–184
  inner voice of, 184–185, 192–193
  language learned by, 162, 165, 181–184, 207, 226,
    228
  learning by, 133–134
  love in, 228–229
  memory control by, 192–193, 194, 207
  photographic memory in, 98, 99
  silent reading by, 205
  social emotion formed by, 224–229
  social rules acquired by, 150
  thinking aloud by, 185
  thought in, 130, 133–134
  see also infants; mother-child relationship
chimpanzees, 13–14, 16, 18, 20, 27, 128
  brain size of, 33–34
  communication of, 145, 146–150, 153, 154–157,
    158, 186
  consciousness of, 89, 178
  emotion in, 146, 186, 214, 224
  food memory of, 86
  food sharing by, 28
  human genetic closeness to, 31, 32, 33, 153, 262
  male dominance in, 146, 147–148
  mother-child relationship among, 28, 148–149,
    152, 157, 158
  pygmy, 17
  reproduction rate of, 23–24
  skills learned by, 41, 132–133, 157, 189
  social behavior of, 146–150, 151, 152–153, 197
  social organization of, 28, 89, 156, 158
  thought in, 129–130, 132–133
  tool use by, 41, 42, 132–133
  as tournament species, 30
  vocal tract of, 159–160
China, 174
  writing of, 205
choking, 160
chunks, 60
  of memory, 91–92
cities, 174, 202
civilization, 174–175, 195, 204–209, 231, 233
  mathematics in, 174, 204–205
  privileged classes of, 174
  river-valley, 174
  South American Indian, 204
  technology of, 207–209
  writing in, 174, 204–208
climate, 15–17
  of ice ages, 9, 21–22, 44, 45, 47, 171, 172, 173,
    174, 204
computers, 140, 232
  consciousness vs., 237–238

decision support systems of, 261
memory and, 193–194, 208, 261
self-awareness and, 260–262
concentration, 243, 246, 260
conscience, 188, 196–198, 200, 203, 226, 252
consciousness, 8, 12, 70, 72, 75, 85–89, 142, 188,
    235–250
  of animals, 86, 87–89, 121, 188
  attention in, see attention
  body metaphors in, 237
  brightness of, 251
  of chimpanzees, 89, 178
  computers vs., 237–238
  emotion in, 215–216, 219–220
  Freudian unconscious vs., 94, 249–250
  inner control of, 89, 185, 211
  nets in, 82–84, 92–93, 177–178, 215, 235–238, 240
  pain in, 219–220
  self-awareness vs., 235–248, 259–260
  working memory and, 92
crows, 36
cultural evolution, 189–209, 213, 214
  antisocial groups in, 150–151, 201–202
  children and, 191–195
  civilization in, 174–175, 195, 204–209, 231, 233
  conscience in, 188, 196–198, 200, 203, 226, 252
  of emotion, 214, 224–233
  fear of outgroups in, 231–232
  genetic equivalents for, 189–190, 206
  life-styles in, 199–200, 202
  memory control required by, 192–195, 203, 207
  mental isolation caused by, 262–263
  modern society in, 150, 199–202, 203, 207–209,
    252, 256, 262
  need to belong and, 201–202, 225
  personal freedom in, 198–202, 262–263
  pleasure/pain continuum in, 224–226
  policing methods in, 197–198, 252
  primitive cultures and, 197, 199–200, 207
  in self-awareness, 202–203, 248, 251–253, 256–
    257, 258, 260–263
  social masks in, 200–202, 203, 252, 258, 260, 263
  social rules in, 195, 196–198
  subcultures in, 199, 200–202, 208
culture, 11, 159, 165
  cold-blooded aggression and, 218
  of Homo sapiens, 172
  language influenced by, 165, 167, 169–170, 183
  love influenced by, 229–230
  male-controlled, 230
  primitive, see primitive cultures
  thought influenced by, 128, 140–141, 209

Dark Ages, 206
Dart, Raymond, 26
deaf people, sign language of, 186–187
decision support systems, 261
déjà vu, 77
dendrites, 38
disgust, 212–214, 227
Disney, Walt, 112
division of labor, 25, 27, 29, 31, 40
dogs, 24, 32–33, 89, 101, 128, 129, 131
dolphins, 129
dreams, 91, 240–241, 249
drugs, mood-altering, 222
ducks, 36

early man, see hominids
education, 99, 166, 261
Egyptians, 205

elephants, fossil, 46
emotion, 8, 10, 48, 49, 76, 87, 210–233, 260
  animal cries expressive of, 146, 186
  in animals, 131–132
  arousal in, 214–218, 219, 220, 223, 226–227
  basic elements of, 214–223
  brain and, 214–216, 217, 219–220, 221–222, 250
  categories of, 213
  cerebral vs. physical, 213
  in consciousness, 215–216, 219–220
  context in, 218, 223
  cultural evolution of, 214, 224–233
  definition of, 131
  evolution of, 210–212, 214, 217, 219–221, 224,
    231–232
  higher, 185, 188, 209, 212, 213, 223, 227–228,
    231–232
  invention of, 213–214, 223, 227
  mixed, 219, 223
  in moods, 221–222, 223
  negative, 228, 231–232
  nets in, 215, 221, 225, 227, 230
  physical effects of, 215–217, 223
  pleasure/pain continuum in, 214, 218–222, 223,
    224–226
  shades of, 212–213, 223
  social, 224–229
  spiritual, 228
English language, 166, 169, 170, 173, 184, 205
engrams, 60
evolution, 9–10, 11, 12–14, 15–48, 87, 131, 168, 186,
    187, 234, 238, 258, 262
  of apes, 15–18, 151
  attention in, 242
  brain-size increase in, 13, 19, 25, 31, 33–40, 103
  climate in, 9, 15–17, 21–22, 44, 45, 47
  cultural, see cultural evolution
  of emotion, 210–212, 214, 217, 219–221, 224,
    231–232
  genes and, 31–33, 38–39, 151–152, 212
  growth rates in, 32–33, 36–37, 38–39
  hunter-gatherers and, 25–29, 43, 45
  of memory, 80, 100, 105–106, 107, 116–117, 133
  physical changes in, 18, 19, 30, 31–37
  reproductive strategies in, 17, 22–25, 31, 34–36,
    37, 151
  social organization in, 18–19, 23, 24, 25–31, 40
  speech adaptations in, 33, 144, 159–161, 162, 170
  speed of, 13
  of vision, 69
  see also hominids; language, origin of
extinction:
  of apes, 17
  of hominids, 20, 22, 34, 44, 47
  of mammals, 22, 47
eyewitness evidence, 118

face, evolution of, 33
farming, 26, 173–174, 204
fear, 216–217, 218, 229, 233
  of outgroups, 231–232
feral children, 40
fire, 27, 40, 45–46, 47, 170
fish, 23, 87, 129, 211
food sharing, 25, 27–29, 31, 40, 224
foresight, 252–253, 258, 260
forests, Miocene, 15–16, 18, 21
fovea, 54
frames of reference, 67, 73
freedom, personal, 198–202, 262–263
Freud, Sigmund, 94, 140, 249

Galilei, Galileo, 106
gathering, 26, 27, 29
gaze-following reflex, 153–154, 162
genes, 31–33, 38–39, 151–152, 212
  ape vs. human, 31–32, 33, 153, 161
  equivalents of, in cultural evolution, 189–190, 206
  speech based in, 162–163, 164–165
gestures, 144, 146–147, 149, 153, 156–157, 167, 168
  inner voice vs., 186–187
gibbons, 16, 18
goldfish, 87
Gombe Stream Reserve, 147, 152–153
gorillas, 16, 17, 20, 23, 145
  diet of, 26–27
  human genetic closeness to, 31
  as tournament species, 30
  vocal tract of, 159–160
grain, cultivation of, 173–174
grammar, 165–170, 187
  children's acquisition of, 183–184
  human-centered, 168, 169, 170
  sentences in, 166–167, 168–170, 183
Greeks, 140, 205, 206
guilt, 196, 198, 212, 213, 227, 228

Hell's Angels, 201–202
hippies, 201–202
Holland, Arnhem Zoo in, 147
hominids, 9, 18–22, 142, 217–218
  apelike heads of, 19
  attention of, 248
  bipedalism of, 18, 19, 22, 24, 25, 33
  brain size of, 18, 19, 20, 34, 43, 44, 47, 163, 171
  conscience of, 196–197
  cultural evolution of, 190–191, 192, 196–197,
    199–200, 202
  diet of, 20–21, 22, 25–27, 33, 44, 46
  emotion in, 223–224
  extinction of, 20, 22, 34, 44, 47
  fire used by, 27, 40, 45–46, 47, 170
  first appearance of, 18
  hands of, 18, 25, 42
  memory in, 107, 120–121
  origin of language and, 150, 153, 154, 156–159,
    160–161, 166, 167, 170–171, 186,190, 192, 203
  predators of, 18, 25
  reproductive strategy of, 24–25, 31, 34, 151
  as savage killers, 26
  self-awareness of, 203, 253
  social organization of, 18, 25–31, 150, 156
  species of, 20–21, 22, 32, 34, 43–47; see also
    Neanderthal man
  thought in, 130–131, 135, 141
  tools of, 27, 34, 40, 42–45, 46, 47, 170
Homo, 20–21, 22, 26, 34, 46, 107, 151, 157, 170
Homo erectus, 34, 44–47, 171, 192
  brain size of, 34, 44, 47, 163
  extinction of, 47
  fire used by, 45–46, 47, 170
  hunting by, 46
  self-awareness of, 203
  speech in, 160–161, 166, 167, 170
  tools of, 44–45, 170
Homo habilis, 46, 153, 154
  brain size of, 34, 43
  tools of, 34, 43–44, 47
Homo sapiens, 34, 47, 204
  culture of, 172
  migration of, 171
  origin of language and, 143, 161, 171–172
Hopi Indians, 169
horses, 36, 39

hunter-gatherers, 25–29, 43, 45, 172, 173, 174, 217–218
hunting, 27–28, 29, 46, 211

ice ages, 9, 21–22, 44, 45, 47, 171, 172, 173, 174, 204
iconic (photographic) memory, 14, 90, 98–99
identity, sense of, 121, 191–192, 194, 234–235, 236, 253–255
illusions, perceptual, 69–70
imagination, 48, 49, 128, 142, 156, 178, 185, 188, 192–193, 232
    language in, 111–112, 117
    limitations of, 112–114, 240
    memory and, 97, 98, 111–114, 117
    nets in, 113–114
    real-life sensations vs., 113–114
India, 39–40, 174
Industrial Revolution, 140
infants, 144, 226
    gaze-following reflex in, 53, 162
    language learned by, 39, 153, 162, 181–182
    learning by, 133–134
    newborn, 37, 38–39, 162
    turn-taking behavior in, 162
    vision of, 38
    vocal tract of, 160
inner voice, 10, 12, 14, 88, 176–209, 225, 247
    of children, 184–185, 192–193
    control over, 178–179
    gestures vs., 186–187
    nature of, 176–177, 178
    reading and, 205–206
    in speech centers, 178–184
    thinking aloud vs., 185–186
    see also cultural evolution
insight, 78, 129, 130, 137
intelligence, 13, 23, 129, 151, 153, 181
    of apes, 17, 24, 25, 26, 145, 146–147, 149, 258

Japan, 42, 157
Java man, 46
Jericho, 174

knuckle-running, 16, 18
K-type reproductive strategy, 22, 23–25, 31, 36, 151
!Kung people, 26, 27–28

language, 9–10, 11, 27, 87, 88, 141, 142, 214, 224
    abstract labels in, 12, 49–50, 212, 213, 223, 225, 234
    in concentration, 247
    cultural influence on, 165, 167, 169–170, 183
    English, 166, 169, 170, 173, 184, 205
    features of, 187
    grammar of, 165–170, 183–184, 187
    in imagination, 111–112, 117
    learning of, 39, 153, 162, 165, 181–184, 207, 226, 228
    memory and, 109–112, 117, 120
    nets and, 155–156, 161, 182, 188–189
    perception affected by, 110–111, 182, 190, 216, 223, 238
    sign, 186–187
    symbolic nature of, 155, 172–173
    in thought, 131, 132
    words in, 111–112, 155, 162, 165, 166–167, 168, 173, 182–183, 213
    see also inner voice; speech
language, origin of, 13–14, 47–48, 143–175, 186, 187–188, 192
    animal communication and, 144–146, 186
    chimpanzee communication and, 145, 146–150, 153, 154–157, 158, 186

civilization and, 174–175
    gestures in, 144, 146–147, 149, 153, 156–157, 167, 168
    hominids and, 150, 153, 154, 156–159, 160–161, 166, 167, 170–171, 186, 190, 192, 203
    Homo sapiens and, 143, 161, 171–172
    protolanguage in, 157–158, 159, 166, 168, 170
    social behavior in, 146–159
    theories of, 143–144
learning, 48, 49, 60–61, 132–134, 142
    by animals, 41–42, 100–103, 132–133, 157, 189
    of language, 39, 153, 162, 165, 181–184, 207, 226, 228
    nets in, 103–104
    relearning vs., 40, 164
leopards, 19, 25, 28, 217
Linguistic Society of France, 143
lions, 149
    hunting by, 27–28
    reproductive strategy of, 23
lizards, 87, 129
logic, 124, 134, 135, 136, 137, 232, 233, 261
    of grammar, 168, 183–184
long-term memory, 90–91, 92, 96–100, 103–120
    chemical changes in, 103–104
    factual, 96, 118–120, 151, 188
    general knowledge in, 105, 106–107, 112, 116, 117, 119, 133–134, 225
    misconceptions about reality and, 106–107
    nets in, 103–105, 119
    personal, 87, 96, 107–110, 114–118, 119–120, 185, 188
    powerful and unique experiences in, 105, 115–116
    purpose of, 105–106
    see also recall; recognition; working (short-term) memory
love, 212, 219, 223, 228–231
    in children, 228–229
    essence of, 230–231
    romantic, 228, 229–230
    in teenagers, 229
Luria, Aleksandr, 96–98
lust, 216, 217, 229, 230

male dominance, 30, 146, 147–148
    cultural, 230
mammals, 41, 147, 211–212
    brains of, 13–14
    emotions of, 211–212, 214, 221
    extinction of, 22, 47
    ice-age, 21–22, 171
    learning by, 41–42
    reproductive strategies of, 23
mammoths, 21
marmosets, 29–30
marsh tits, 86
mathematics, 135–136, 174, 204–205, 261
memory, 8, 9, 10, 12, 48, 49, 85–123, 142, 156, 178, 190, 232
    in animals, 86–87, 95–96, 100–103, 105–106, 110, 111, 116–117, 122
    association of ideas in, 81, 95–96, 108, 109–110, 127
    brain damage and, 90–91
    chunks of, 91–92
    classification by, 80–81
    computers and, 193–194, 208, 261
    control of, 121–122, 132, 192–195, 203, 207, 227
    definition of, 79–80
    of dreams, 91
    evolution of, 80, 100, 105–106, 107, 116–117, 133
    extraordinary, 96–99, 117
    eyewitness evidence and, 118

forgetting in, 118
imagination and, 97, 98, 111–114, 117
language and, 109–112, 117, 120
limitations of, 80–82, 90–91, 112–117
nets and, 60, 61, 62–68, 72–82, 83, 254
photographic, 14, 90, 98–99
pleasure/pain continuum and, 221
in self-awareness, 121–123
synesthesia in, 97
techniques for improving, 193
thought and, 132–134, 139
unconscious, 94, 249–250
see also learning; long-term memory; working
    (short-term) memory
Middle Ages, 140, 205
Middle East, 143, 173, 174, 204
mind, 9–14, 49–50, 232, 236–237, 262
    artificial abilities of, 187–188
    basic components of, 10, 12, 48, 49
    uncluttered, 88, 99
Miocene era, 15–18, 21, 33
monkeys, 58, 214
    baboons, 19, 27, 30, 41, 87, 189
    communication of, 144–145, 150
    skills learned by, 41–42, 157, 189
    social organization of, 18–19, 29–30
    vervet, 40, 144–145
monogamy, 29–31
moods, 221–222, 223
mother-child relationship, 133, 144, 157–158, 191–
    192, 228
    among chimpanzees, 28, 148–149, 152, 157, 158
    food sharing in, 28
    language learning in, 153, 162, 181–182, 226
    social emotion and, 226
mouth-breeding fish, 23
myelinization, 37–40, 164–165

Navaho Indians, 169–170
navigation, 134–135
Neanderthal man, 170–171
    brain size of, 171
    inner voice of, 176, 185–186
    speech of, 171, 176
    vocal tract of, 171
nets, 49–84, 141–142, 213
    in attention, 241, 242–244, 245–247
    boundaries lacking in, 61, 82
    in consciousness, 82–84, 92–93, 177–178, 215,
        235–238, 240
    description of, 60–65
    of dreams, 240–241
    in emotion, 215, 221, 225, 227, 230
    experience and, 62–64
    flowing permanence lent by, 83
    in imagination, 113–114
    language and, 155–156, 161, 182, 188–189
    in learning, 103–104
    in long-term memory, 103–105, 119
    memory and, 60, 61, 62–68, 72–82, 83, 92–93, 96,
        103–105, 119, 253–254
    movement of, 62
    perception and, 68–73, 83
    in self-awareness, 236, 254–256
    short-lived character of, 61
    thought and, 127–128, 130, 133, 137–139
    vision and, 51–60, 63, 68–72
    in working memory, 92–95, 96
neurons, 50–51, 79–80
    dendrites of, 38
    junctions of, 102–104
    myelinization of, 37–40, 164–165
    see also nets

Nigeria, 98–99
noradrenaline, 217

orangutans, 16, 17, 23, 145
    brain size of, 33–34
oysters, 23

pair bonding, 25, 30–31, 40
parrots, 34, 36
passions, 216–218
Pavlov, Ivan, 102
Peking man, 46
perception, 49, 92, 142, 251
    in animals, 85, 106, 111
    definition of, 72
    general vs. detailed, 72
    illusions of, 69–70
    language and, 110–111, 182, 190, 216, 223, 238
    nets and, 68–73, 83
    working memory and, 94–95
    see also vision
photographic (iconic) memory, 14, 90, 98–99
pictographic writing, 204–205
pigeons, 87
Plato, 140, 205
pleasure/pain continuum, 214, 218–222, 223, 224–
    226
predators, 18, 19, 25, 217
primitive cultures, 45, 134–135, 136, 169–170
    cultural evolution and, 197, 199–200, 207
    see also hunter-gatherers
processing frames, 60
punks, 201–202
pygmies, Congo, 26, 29, 45
pygmy chimpanzees, 17

rabbits, 23, 24, 36, 211, 242
rage, 212, 216–217, 218, 233
reading, 164, 187, 207
    silently vs. aloud, 205–206
recall, 74, 75, 90, 107–110, 114–115, 121, 180
    mental strategies used for, 108–110
recognition, 74–78, 84, 100, 105–106, 117, 119, 243–
    244, 245–246, 254–256
    "aha" feeling of, 75–77, 78, 107, 130, 177
    by animals, 86
    déjà vu in, 77
    recall vs., 74, 75, 107–108, 109
    testing of, 74
    wishful thinking in, 78
reproductive strategies, 22–25
    of apes, 17, 22, 23–25, 36, 37, 151
    of birds, 34–36
    of hominids, 24–25, 31, 34, 151
    human, 24–25
    K-type, 22, 23–25, 31, 36, 151
    overlapping of children in, 25
    r-type, 22–23, 24
retina, 52–54, 56, 58, 113
    fovea of, 54
ridicule, 197
Roman Catholic Church, 198
Romans, 204–205, 206
romantic love, 228, 229–230
r-type reproductive strategy, 22–23, 24
runner's high, 222

saber-toothed tigers, 21
salamanders, 32
schemata, 60
science, 134, 135–136, 262
    see also technology
scripts, 60

sea slugs (Aplysia), 100–103, 210
self-awareness, 8, 9, 10, 12, 48, 49, 61, 87, 88, 141,
     185, 188, 232, 234–263
   antisocial behavior vs., 251–252, 253
   of brain, 121–122
   consciousness vs., 235–246, 259–260
   cultural evolution in, 202–203, 248, 251–253,
     256–257, 258, 260–263
   definition of, 121
   in foresight, 252–253, 258, 260
   future of, 258, 260–263
   improvement of, 258–260, 262
   limitations of, 248–251
   memory in, 121–123
   metaphorical distance in, 126
   nets in, 236, 254–256
   physical actions vs., 122–123
   self-obsession vs., 251
   sense of identity in, 121, 191–192, 194, 234–235,
     236, 253–255
   in social masks, 252, 258, 260, 263
   technology and, 260–262, 263
sexual signals, 30, 144
Shereshevskii, 96–98, 117, 193
short-term memory, see working (short-term) mem-
     ory
sign language, 186–187
slings, 43
social behavior, 146–159, 262
   of chimpanzees, 146–150, 151, 152–153, 197
   social rules in, 150–151, 195, 196–198, 227
   see also antisocial behavior
social emotion, 224–229
social masks, 200–202, 203, 252, 258, 260, 263
social organization, 25–31, 150
   of apes, 18, 24, 29–30
   of chimpanzees, 28, 89, 156, 158
   division of labor in, 25, 27, 29, 31, 40
   food sharing in, 25, 27–29, 31, 40, 224
   of monkeys, 18–19, 29–30
   pair bonding in, 25, 29–31, 40
South American Indians, civilization of, 204
Spain, 46
spears, 42–43
speech, 47–48, 107, 159–165, 187, 203
   arched roof of mouth in, 160–161, 171
   genetic basis of, 162–163, 164–165
   of Homo erectus, 160–161, 166, 167, 170
   of Neanderthal man, 171, 176
   physical adaptation to, 33, 144, 159–161, 162, 170
   speed of, 161–162
   vocal tract in, 159–161, 162, 171
   in working memory, 161–162, 169
   see also language
speech centers, 39–40, 162–163
   inner voice in, 178–184
   sentence formation in, 179–181
strokes, 40, 91, 163–164
synesthesia, 97

Tanzania, 45
technology, 207–209, 260–262, 263
   computer, see computers
teenagers, love and, 229
termites, 41
thinking aloud, 185–186
thirst, 222
thought, 8, 48, 49, 61, 87, 95, 111, 123, 124–142,
     188, 203, 216, 218, 249
   abstract, 141

   in animals, 128–130, 131–133, 138
   cultural influence on, 128, 140–141, 209
   definition of, 124
   inner control of, 89, 131, 138, 192
   intuitive, 134–135
   language in, 131, 132
   logical, 124, 134, 135, 136, 137, 232, 233, 261
   mechanisms of, 124–128
   memory and, 132–134, 139
   metaphors in, 138–141
   natural, 130, 132–135
   nets and, 127–128, 130, 133, 137–139
   rational, 124, 134, 135–138
   writing and, 207
tigers, 217
   saber-toothed, 21
time, 92–94, 234
   linguistic expression of, 169
tools, 27, 34, 40–45, 46, 47, 132–133, 170
touch, sense of, 55–56
   illusions in, 70
tournament species, 30
Truk Islanders, 134–135, 136

Uganda, 40
unconscious, Freudian, 94, 249

vervet monkeys, 40, 144–145
Victorian age, 19
vision, 51–60, 64, 68–70, 94, 95, 103–104, 113–114,
     188
   blind spot in, 53–54
   camera compared with, 52
   contrasts sharpened in, 52–53, 72, 238, 246
   depth perception in, 53, 56
   distance judgment in, 69
   eyeball in, 52, 54, 56
   feature detectors in, 53
   focus in, 54, 58
   illusions of, 69–70
   in infancy, 38
   movement compensations of, 56, 69
   optic nerves in, 54–55, 56
   reeducation of, 70
   retina in, 52–54, 56, 58, 113
   visual cortex in, 56–60, 61, 65, 69, 73, 75, 82, 112
vocal tract, 159–161, 162, 171
voice box, 144, 160

words, 111–112, 155, 162, 165, 166–167, 168, 173,
     182–183, 213
working (short-term) memory, 90–96, 104, 108, 112,
     115, 137
   association of ideas in, 95–96
   attention and, 93–95, 244
   boundaries lacking in, 92
   capacity of, 91–92
   duration of, 90, 92–93
   education and, 99
   nets in, 92–95, 96
   perception and, 94–95
   reading in, 206
   speech in, 161–162, 169
   time and, 92–94
   see also long-term memory
writing, 164, 166, 174, 204–208
   modern media of, 207–208, 261
   pictographic, 204–205

Yo-Heave-Ho theory, 143–144